AA004930

XIII Bulgarian-Serbian Astronomical Conference (XIII BSAC)

Publications of the Astronomical Society "Rudjer Boskovic"
Number 25

Velingrad, Bulgaria
3-7 October 2022

Editors:

Evgeni Semkov
Milan S. Dimitrijević
Momchil Dechev
Zoran Simić

ISBN: 978-1-7138-7543-7

Printed from e-media with permission by:

Curran Associates, Inc.
57 Morehouse Lane
Red Hook, NY 12571

Some format issues inherent in the e-media version may also appear in this print version.

Copyright© (2022) by Astronomical Society "Rudjer Bošković"
All rights reserved.

Printed with permission by Curran Associates, Inc. (2023)

For permission requests, please contact Serbian Astronomical Society
at the address below.

Astronomical Society "Rudjer Bošković"
Kalemegdan
Gornji Grad 16
11000 Belgrade
Serbia

Additional copies of this publication are available from:

Curran Associates, Inc.
57 Morehouse Lane
Red Hook, NY 12571 USA
Phone: 845-758-0400
Fax: 845-758-2633
Email: curran@proceedings.com
Web: www.proceedings.com

ПУБЛИКАЦИЈЕ АСТРОНОМСКОГ ДРУШТВА "РУЂЕР БОШКОВИЋ" Св. 25
PUBLICATIONS OF THE ASTRONOMICAL SOCIETY "RUDJER BOŠKOVIĆ" No. 25

PROCEEDINGS
OF THE XIII BULGARIAN-SERBIAN ASTRONOMICAL CONFERENCE

Velingrad, Bulgaria, October 3-7, 2022

Eds. Evgeni Semkov, Milan S. Dimitrijević, Momchil Dechev and Zoran Simić

BELGRADE, 2023

Participants of the 13th Bulgarian-Serbian Astronomical Conference held in Velingrad (Bulgaria) from 3-7 October 2023.
In front: Vladimir Božilov, Todor Veltchev, Luba Dankova, Orlin Stanchev
First row standing: Evgeni Semkov, Yavor Chapanov, Tsvetan Georgiev, Milan S. Dimitrijević, Stefan Stefanov, Milen Minev, Anatoly Miroshnichenko, Krasimira Yankova, Tanyu Bonev
Second row: Dejan Urošević, Zoran Simić, Alok Gupta, Rositsa Miteva, Galin Borisov, Dragomir Marchev, Renada Konstantinova-Antova, Milica Vučetić
Third row: Stanislav Milošević, Miljana Jovanović, Momchil Dechev, Daniela Kirilova, Sava Donkov, Nataliya Pavlova, Rumiana Bojilova, Ognyan Kounchev, Bojan Arbutina
Fourth row: Boyko Mihov, Svetoslav Botev, Jakub Tokarek, Daniela Boneva, Magdalena Christova, Tsvetan Tsvetkov, Goran Damljanović, Elena Vchkova-Bebekovska, Krystian Ilkiewicz, Svetoslav Hristov, Luba Slavcheva-Mihova, Lubomir Iliev, Asen Mutafov
Behind: Kiril Stoyanov, Vladislav Marchev, Rumen Bachev

ПУБЛИКАЦИЈЕ АСТРОНОМСКОГ ДРУШТВА "РУЂЕР БОШКОВИЋ"
PUBLICATIONS OF THE ASTRONOMICAL SOCIETY "RUDJER BOŠKOVIĆ"

Св. 25 No. 25

PROCEEDINGS OF THE XIII BULGARIAN-SERBIAN ASTRONOMICAL CONFERENCE

Velingrad, Bulgaria, October 3-7, 2022

Eds. Evgeni Semkov, Milan S. Dimitrijević,

Momchil Dechev and Zoran Simić

Б Е О Г Р А Д
2023

PUBL. ASTRON. SOC. "RUDJER BOŠKOVIĆ" No. 25, 1-200 BELGRADE, APRIL 2023

SCIENTIFIC COMMITTEE

Evgeni **Semkov** (IANAO) - chairperson
Luka Č. **Popović** (AOB) - co-chairperson

Renada **Konstantinova-Antova** (IANAO)
Lubomir **Iliev** (IANAO)
Kamen **Kozarev** (IANAO)
Dragomir **Marchev** (ShU)
Kostadinka **Koleva** (SRTI)
Petko **Nedialkov** (SU)
Rumen **Bachev** (IANAO)
Nikola **Petrov** (IANAO)
Luba **Slavcheva-Mihova** (IANAO)
Ognyan **Kounchev** (IMI)
Zoran **Simić** (AOB)
Nenad **Sakan** (IPhB)
Rade **Pavlović** (AOB)
Milan **Stojanović** (AOB)
Oliver **Vince** (AOB)
Dejan **Urošević** (UB)
Andjelka **Kovačević** (UB)
Bojan **Arbutina** (UB)
Milica **Vučetić** (UB)
Gordana **Apostolovksa** (US)
Milan S. **Dimitrijević** (Member Emeritus)
Milcho **Tsvetkov** (Member Emeritus)

LOCAL ORGANIZING COMMITTEE

Evgeni **Semkov** (IANAO)
Georgi **Latev** (IANAO)
Momchil **Dechev** (IANAO)
Dimitar **Sotirov** (IANAO)
Luba **Dankova** (IANAO)
Asen **Mutafov** (IANAO)

Legend:
IANAO - Institute of Astronomy with National Astronomical Observatory, Bulgarian Academy of Sciences
AOB - Astronomical Observatory, Belgrade
ShU - Shumen University
SRTI - Space Research and Technology Institute, Bulgarian Academy of Sciences
IPhB - Institute of Physics, Belgrade
SU - Sofia University
UB - University of Belgrade
IMI - Institute of Mathematics and Informatics, Bulgarian Academy of Sciences
US - Ss. Cyril and Methodius University in Skopje

ORGANIZER:
Institute of Astronomy and National Astronomical Observatory,
Bulgarian Academy of Sciences, Sofia, Bulgaria

Co-organizer:
Astronomical Observatory, Belgrade, Serbia

Text arrangement by computer: Tatjana Milovanov

Published and copyright © by Astronomical Society "Rudjer Bošković", Kalemegdan, Gornji Grad 16, 11000 Belgrade, Serbia
President of the Astronomical Society "Rudjer Bošković": Miodrag Dačić

Production: Skripta Internacional, Mike Alasa 54, Beograd, in 100 copies

CONTENTS

Proceedings of the XIII Bulgarian-Serbian Astronomical Conference (XIII BSAC) Velingrad, Bulgaria, October 3-7, 2022
Editors: Evgeni Semkov, Milan S. Dimitrijević, Momchil Dechev and Zoran Simić
Publ. Astron. Soc. "Rudjer Bošković" No 25, 2023, 5-9

LATEST ACHIEVEMENTS FROM BELOGRADCHIK OBSERVATORY

RUMEN BACHEV

Institute of Astronomy and NAO, Bulgarian Academy of Sciences
Tsarigradsko Shose Blvd. 72, Sofia 1784, Bulgaria
E-mail: bachevr@astro.bas.bg

Abstract. We present some of the resent (since Sept. 2020) instrumental developments at Belogradchik Astronomical Observatory, Bulgaria, and show examples of the first results. More specifically these include the first blazar polarization results, obtained with our new filter-based linear polarimeter, as well as the first transient event identification attempt, with the new wide-field camera.

1. INTRODUCTION: BELOGRADCHIK OBSERVATORY

Belogradchik Astronomical Observatory is located at the NW part of Bulgaria, near the town of Belogradchik (at the northern outskirts of the Balkan mountain) at 610m altitude. The Observatory has been built in 1965 and since 1969 has operated a 60cm equatorial Cassegrain telescope, the largest instrument of its class in the Balkan peninsula at the time it was mounted. Throughout the first years the telescope has been equipped with electrophotometers as light detectors, but during the last 20 years mostly CCD's, equipped with standard UBVRI filters have been used (e.g. Strigachev & Bachev, 2011).

2. POLARIMETRIC MEASUREMENTS

Linear polarization measurements at Belogradchik Observatory were initiated mostly for purposes to study blazars. Blazars are a type of active galactic nuclei, in which most of the optical emission comes from a jet that happens to be pointed in the direction of the observer (Urry & Padovani, 1995). This emission is produced via synchrotron processes by relativistic particles in magnetic field and is linearly polarized (e.g. Rybicki & Lightman, 1986).

We used polarization filters, mounted on a double-barrel (7x7) filter set, so the polarization filters could be combined with any of the other photometric filters

(UBVRI). Instead of the normally used 4 polarimetric filters (oriented at 0, 45, 90, 135 degrees), our setup included only 3 filters, oriented at 0, 60 and 120 degrees. Still, the polarization parameters of the source can be derived, however the calculation is somehow more difficult, as the standard Stokes parameters, which would significantly facilitate the calculations, this time cannot be used.

Thus, if consider the light from a polarized source to consist of entirely polarized (I_p) and non-polarized (I_{np}) parts, then the measured intensities through the 3 polarization filters will be:

$$I_0 = \frac{1}{2} I_{np} + I_p . cos^2\theta$$

$$I_{60} = \frac{1}{2} I_{np} + I_p . cos^2(\theta - 60)$$

$$I_{120} = \frac{1}{2} I_{np} + I_p . cos^2(\theta - 120)$$

We solve numerically the equation above for I_{np} and I_p. To assess the errors of these values, the equations have been solved repeatedly, varying the measured intensities within their respective measurement errors. Then for the polarization rate and the orientation (electric vector polarization angle, *EVPA*) one gets:

$$p[percent] = 100 \frac{I_p}{I_p + I_{np}}$$

$$EVPA \equiv \theta$$

As of the end of 2022 more than 20 blazars have been monitored polarimetrically, some of which on the intra-night time scale (e.g. Bachev, 2022a, 2022b, 2022c; Bachev & Strigachev, 2022). Some of our polarimetric measurements have already been published in major international journals even of the rang of *Nature* (Jorstad et al., 2022). Many of these objects showed significant changes and high polarization rates. For example, the blazar S4 0954+65 showed polarization rate of about 39% during the night of June 5, 2022 (Bachev, 2022a).

One of the most polarimetrically studied objects was BL Lacertae, the prototype of the blazar family. For this object, during its unprecedented 2020-2021 outburst we detected significant intra-night changes, not only in flux but also in polarization (Bachev et al., 2022). Figure 1 presents some of the results.

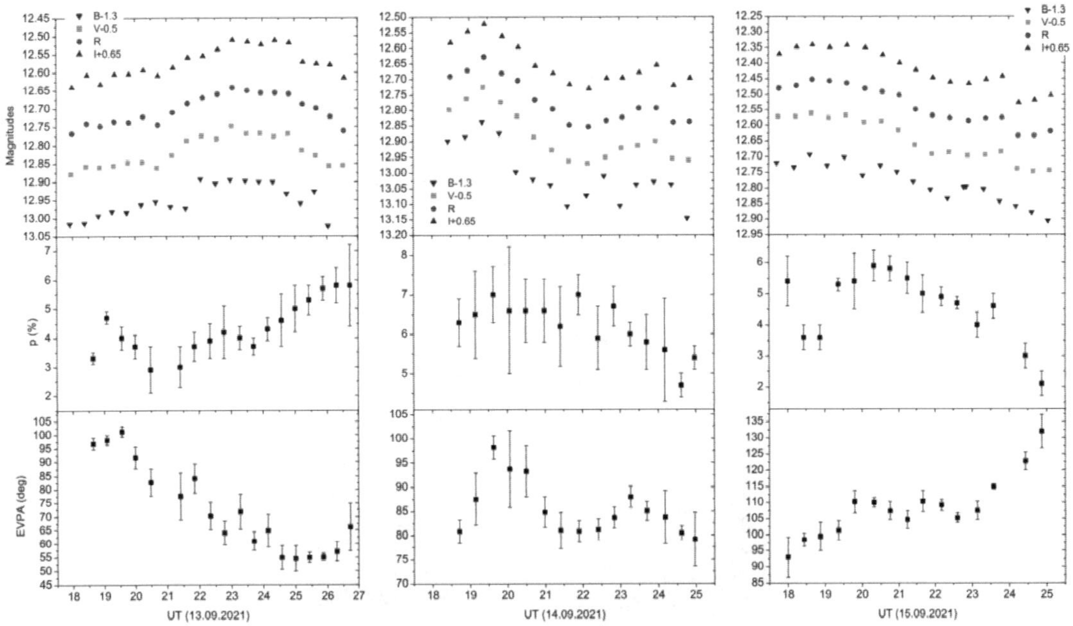

Figure 1: Intra-night multicolor and R-band polarimetric observations with the 60cm Belogradchik telescope. BVRI light curves (top panel) are shown together with polarization rate (middle panel) and EVPA (bottom panel) for each night. Significant variability in both flux and polarimetric parameters can be seen (Bachev et al., 2022).

3. WIDE-FIELD IMAGING

Another addition to the Belogradchik Observatory equipment is recently attached to the 60cm telescope an 8” F/2 *Celestron* Rowe-Ackermann Schmidt Astrograph for wide-field (around 3 degrees with the used *ZWO* ASI071MC CMOS) imaging, Figure 2. This instrument is intended mostly for fast optical identifications of transient events (GRB, neutrino or GW events, etc.), whose location is known only approximately. The idea is that if a transient, detected in other wavelengths or by other multi-messenger detectors would have an optical counterpart, we might be able to detect it and thus identify the source. Other activities may include comet imaging, educational purposes, etc.

One of the first attempts to identify an optical counterpart of a transient event was a gamma-ray burst (GRB201020B), detected by *Fermi* satellite (Fermi GBM Team, 2020) and located with a 3-degree error. *Master* network detected shortly after the *Fermi* trigger a 13.7 mag object in the field (Lipunov et al., 2020) and thus the object was identified. Unfortunately, our observations could have taken place only about 4 hours later, when the object apparently had faded (Figure 3).

Figure 2: The 8" F/2 Rowe-Ackermann Schmidt Astrograph for wide-field imaging and transient identification, attached to the 60cm telescope.

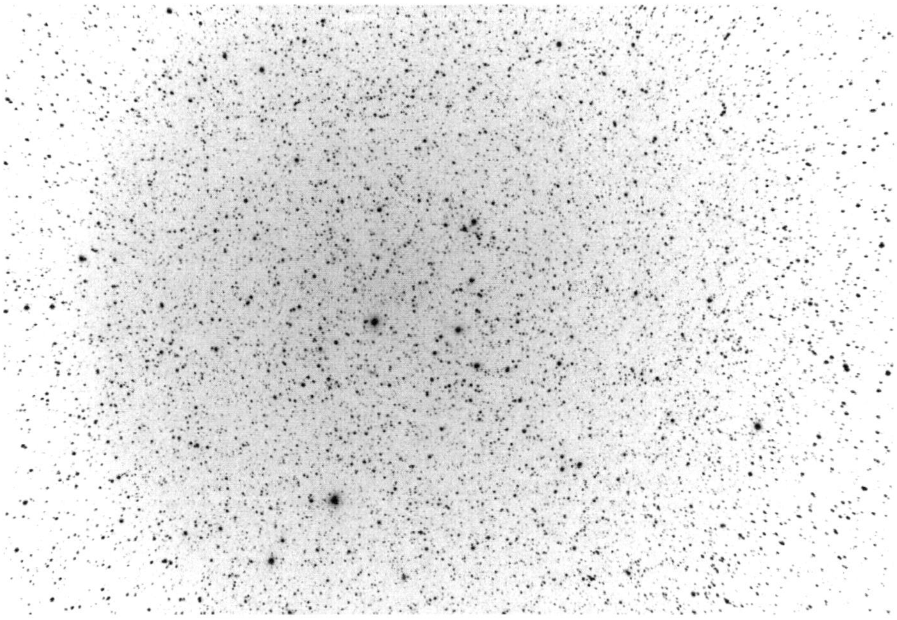

Figure 3: The 3-degree field of the GRB201020B event, observed with the 8" F/2 astrograph. About 4 hours after the Fermi trigger the optical counterpart has already faded and is not seen in the frame.

4. CONCLUSIONS

With these new capabilities to perform optical polarimetry and wide-field imaging we aim to place Belogradchik Observatory among the small, but actively participating in wide international campaigns observatories. Nowadays, a small observatory with a generally modest equipment, located not at an excellent for astronomical observations conditions, can effectively contribute to astrophysical science only via wide co-operation with other similar astrophysical centers. We believe that the new additions at the Belogradchik Observatory, as well as other planned in the near future will help to retain our position of a modern astrophysical center throughout the 21st century.

Acknowledgements

This research was partially supported by the Bulgarian National Science Fund of the Ministry of Education and Science under grants KP-06-H28/3 (2018), KP-06-H38/4 (2019) and KP-06-KITAJ/2 (2020).

References

Bachev R.: 2022a, *ATel* #15414.
Bachev R.: 2022b, *ATel* #15447.
Bachev R.: 2022c, *ATel* #15557.
Bachev R., Strigachev A.: 2022, *ATel* #15220.
Bachev R., et al.: 2022, to be published.
Fermi GBM Team: 2020; *GCN Circular* #28702.
Jorstad S. G., Marscher A. P., Raiteri C. M, et al.: 2022, *Nature*, 269, 465.
Lipunov et al.: 2020, *ATel* #14107.
Rybicki G. B., Lightman A. P.: 1986, "*Radiative Processes in Astrophysics*", Wiley.
Strigachev A., Bachev R.: 2011, *Bulgarian AJ* 16, 144.
Urry C.M., Padovani P.: 1995, *PASP*, 107, 803.

Proceedings of the XIII Bulgarian-Serbian Astronomical Conference
(XIII BSAC) Velingrad, Bulgaria, October 3-7, 2022
Editors: Evgeni Semkov, Milan S. Dimitrijević, Momchil Dechev and Zoran Simić
Publ. Astron. Soc. "Rudjer Bošković" No 25, 2023, 11-16

NEGATIVE SUPERHUMPS IN CATACLYSMIC VARIABLES

SVETLANA BOEVA, GEORGI LATEV,
STEFAN STEFANOV and MIROSLAV MOYSEEV

Institute of Astronomy and NAO, Bulgarian Academy of Sciences
72 Tsarigradsko shosse, 1784, Sofia, Bulgaria
E-mail: sboeva@astro.bas.bg

Abstract. We calculated relations between the known periods of positive and negative superhumps and the orbital periods (Stolz–Schoembs relation) for nova-like variables. Using the linear approximations we obtained slopes around 1.31 and 0.93 for (P_{sh+} - P_{orb}) and (P_{sh-} - P_{orb}) respectively.

In the typical for NLs range of orbital periods 0.1- 0.2 days the corresponding coefficients are 1.07 and 0.91.

1. INTRODUCTION

A common feature in different types of cataclysmic variables (CVs) such as Dwarf Novae (DN), AM CVn, Old Novae, Nova-likes (NLs), is the existence of superhumps (SHs) - periodic changes in brightness with periods close to the orbital period (P_{orb}). If this period is few percents longer than the P_{orb}, then we have positive superhumps (P+), while if variations are with period few percents shorter than the orbital period Porb, we have negative superhumps (P-).

There are many reviews and studies on this phenomena – for example the works of Patterson et al. (2005; 2002), Kato (2022), Montgomery (2009), Bruch (2022), etc. A major difficulty in such studies is the obtaining of long light curves of the objects during several consequent nights. The new space telescopes and missions, like TESS (Transiting Exoplanet Survey Satellite) can help by providing such data for superhumps detections and studies.

Many works show evolution of SHs in DN - during superoutbursts and during normal outbursts, the SH amplitudes and periods are changing. In NLs SHs are stable for months and even years and they are called "permanent" superhums. But in this type of systems, a disappearing of the SHs is observed, a switching from states with positive to negative SHs or even simultaneous existing of these two periods.

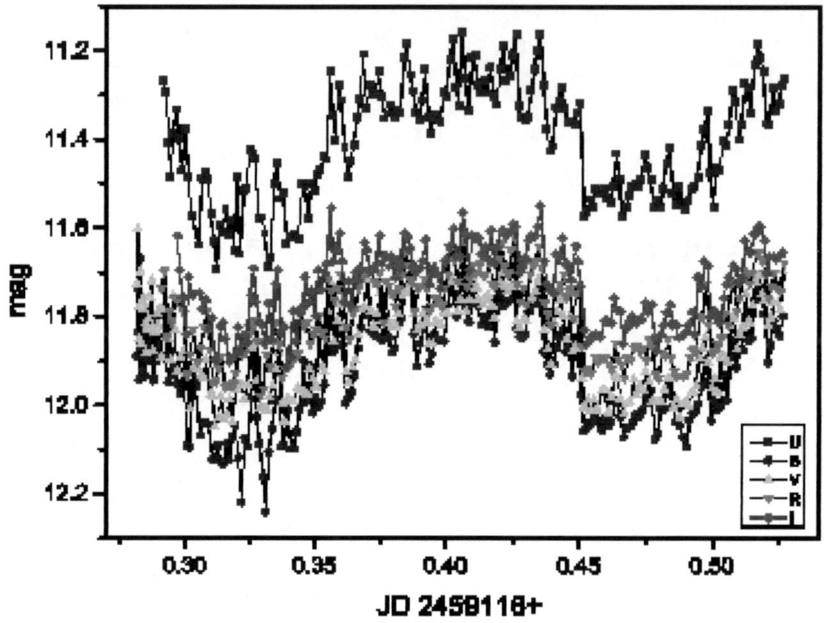

Figure 1: Light curve of BG Tri on September 23rd, 2020. Clearly is visible an existence of negative superhump with period 5 % shorter than the orbital one.

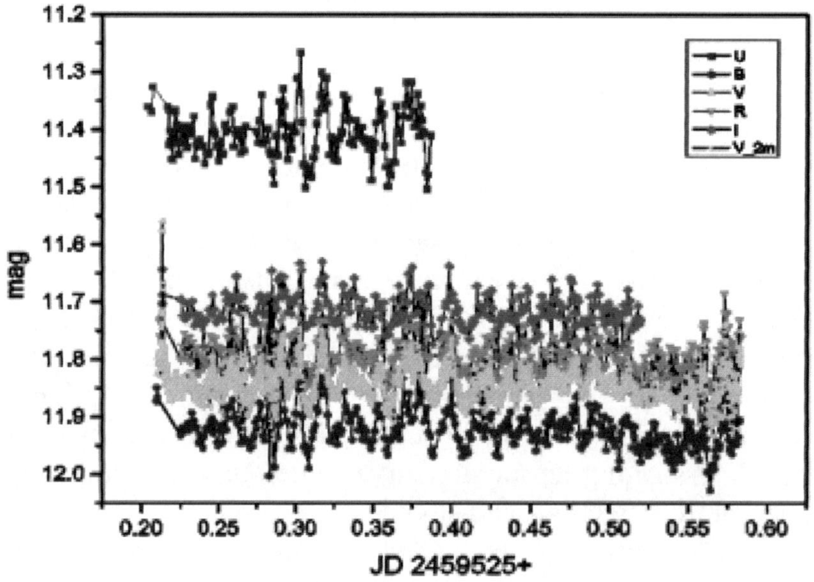

Figure 2: Light curve of BG Tri on November 06th, 2021 – the negative superhump is missing.

Fig.1 and Fig. 2 are example for above mentioned occurrence – the detected in 2020 negative SH in nova-like BG Tri is missing in the light curve from 2021 (Stefanov et al., 2022).

The theory interprets the positive SHs as prograde apsidal precession of an eccentric accretion disk caused by the tidal instability driven by a 3:1 resonance. The most accepted explanation of the nature of negative SHs is nodal retrograde precession of a tilted disk with respect to the orbital plane.

Stolz and Schoembs (1984) proposed a linear relation between P_{sh} and P_{orb}. This relation can be used for determination of P_{orb} using P_{sh} or for an estimation of mass ratio q ($q=M_2/M_1$) of the systems using period excess. Fuentes Morales et al. (2018) published actual linear fit for positive superhumps for several types of CVs. In this work we calculate linear fits for the positive and the negative SHs of NLs.

2. DATA

In the last years, the amount of detected SHs in NLs increased. In 2018 Fuentes Morales et al. have used for the calculation of their (P_{sh}-P_{orb}) linear fit 17 NL systems. In the literature we found 22 NLs with P+ and 26 with P- (Ritter & Kolb, 2005; Wood et al., 2009; Pavlenko et al., 2020; Castro Segura et al., 2021; Iłkiewicz et al., 2021; Bruch, 2022; Stefanov et al., 2022; Stefanov, 2022; Kato, 2022, etc.). Our sample for positive SHs contains all these 17 systems except BK Lyn (probably very old nova) and 6 additional ones. The average excess of the positive superhump period is ~ 9 % and the deficit for negative superhumps is ~ 4%.

10 of the included stars have both types of superhumps: BH Lyn, TT Ari, MV Lyr, PX And, TV Col, AQ Men, LS Cam, BG Tri, V 1193 Ori, V 795 Her. It is not clear why and when they switch the regime of the humps.

There are observations that show slight changes of the periods of the "permanent" SHs, their shapes, amplitudes and colors, but it does not significantly change the relations. Probably they are caused by long-term variations in the accretion disc or some of its structures. Part of NL systems belong to the sub-class VY Scl which sometimes show low photometric states. The decrease of mass-transfer rate leads to disappearance of SHs.

3. RESULTS AND DISSCUSION

We preferred to use Stolz–Schoembs relation from Gänsicke et al. (2009) and Fuentes Morales et al. (2018) in the reversed form

$$P_{sh} = a + b^* P_{orb}.$$

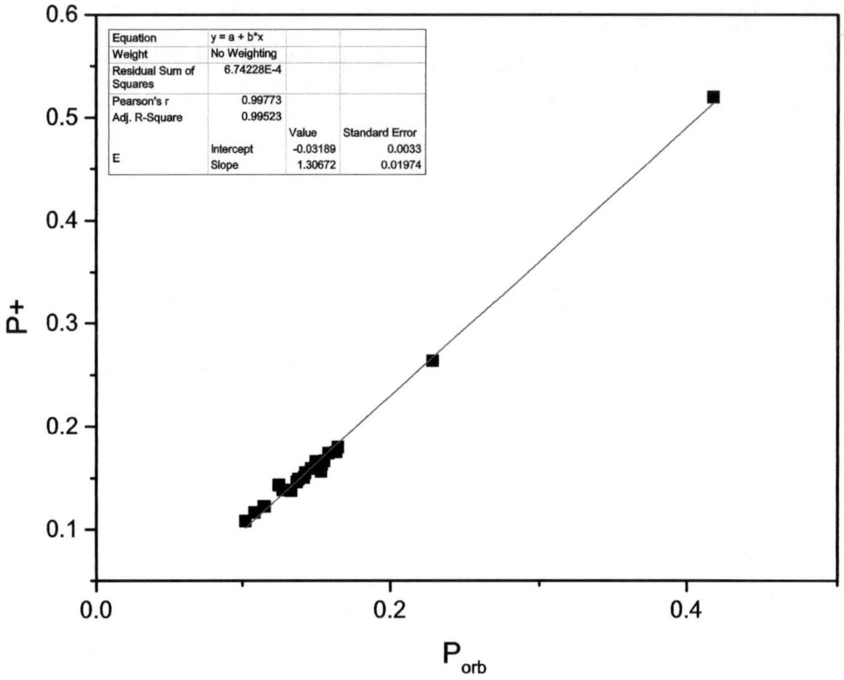

Figure 3: P+ as a function of Porb. Linear fit for 22 NLs.

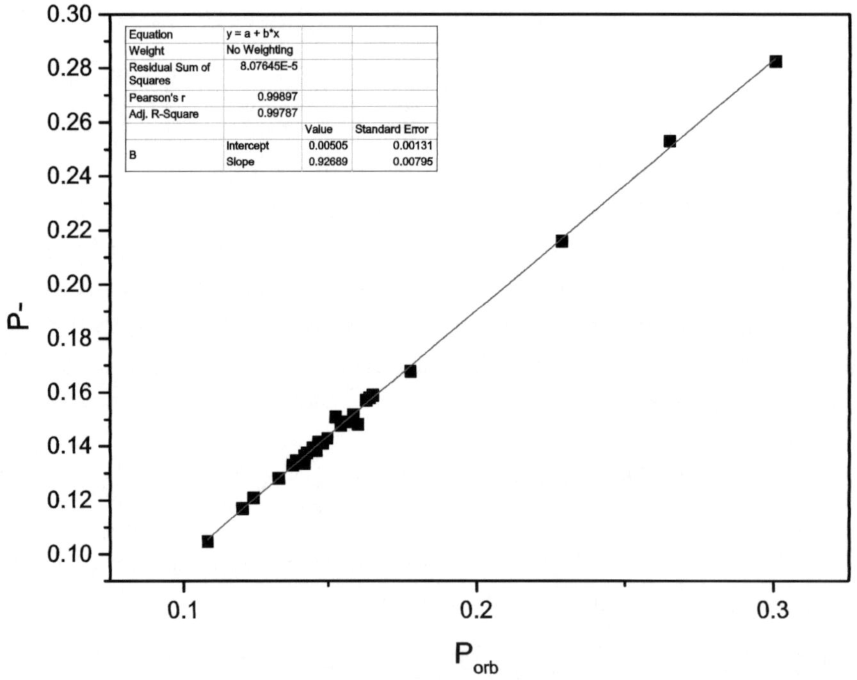

Figure 4: P- as a function of Porb. Linear fit for 26 NLs.

This equation form is more appropriate because we calculated not only the relation for the positive superhumps but also for the negative ones. The corresponding coefficients can easily be converted into one another. In that way we obtained the dependence of the period of the superhump as a function of the orbital period.

We obtained the coefficients of the linear fit as per below:

$$P+ = -0.032\ (\pm\ 0.003) + 1.306\ (\pm\ 0.020) * P_{orb}$$
$$P- = 0.005\ (\pm\ 0.001) + 0.927\ (\pm\ 0.007) * P_{orb},$$

where the periods are given in days (Fig. 3 and Fig.4). Both Pearson correlation coefficients are 0.998 and 0.999, respectively.

If we use the standard form of the equation, then the calculated coefficient of the slope for P+ is 0.76. Gänsicke et al. (2009) in their work used various types of CVs with positive SHs, but Fuentes Morales et al. (2018) determined separately for nova-like variables the slope 0.84 and 0.90 for dwarf novae - nova-like stars have smaller slopes. Our result is slightly less than this value. If we use only systems with typical for NLs orbital periods in the range of 0.1 – 0.2 days, the obtained slope is 0.90 (in reversed relation – 1.07). It seems, the outliers – TV Col and RZ Gru, have changed the coefficient of this relation. Although the dependence gives a good linear fit over a wide interval of periods, probably there exists some difference in all sections. It seems the (P_{sh+} - P_{orb}) relation is more flat for systems with longer orbital periods.

For negative superhumps the slope is 0.93. A recalculation using stars with orbital periods in the interval 0.1 – 0.2 days, obtained value is 0.91 and it is in the standard error interval of the slope. Therefore negative superhumps are more often detected in NLs and probably the linear dependence between their periods and the orbital ones is valid at more wide range.

The relation between P+ and P- for all 10 systems with both types of superhumps gives 1.35 ± 0.04.

4. CONCLUSIONS

We calculated linear fits for Stolz–Schoembs relation for period of SH (P_{sh}) as function of the orbital period (P_{orb}).

The obtained value of the slope for the positive superhumps is 1.31. For systems in the range of orbital periods 0.1- 0.2 days respectively this value is 1.07. It is possible that the relation is more flat for systems with longer orbital periods.

For positive superhumps the slope of the linear fit is 0.93 and it changes slight (0.91) for systems with orbital periods 0.1 – 0.2 days.

The periods of positive superhumps are approximately 35 % longer than the periods of negative ones for systems which have both types.

Acknowledgements

This work is partially supported by the grant КП-06-28/2 “Binary stars with compact object” (Bulgarian National Science Fund).

References

Bruch A.: 2019, *MNRAS,* 489(2), 2961-2975.
Bruch A.: 2022, *MNRAS,* 514(4), 4718-4735.
Fuentes Morales I. et al.: 2018, *MNRAS,* 474(2), 2493-2501.
Gänsicke B. T., et al.: 2009, MNRAS, 397(4), 2170-2188.
Iłkiewicz K., et al.: 2021, *MNRAS,* 503(3), 4050-4060.
Kato T.: 2022, *VSOLJ Variable Star Bulletin,* No. 89.
Kato T.: 2022, *VSOLJ Variable Star Bulletin,* No. 101.
Montgomery M. M.: 2009, *MNRAS,* 394(4), 1897-1907.
Patterson J. et al.: 2002, *PASP,* 114, I. 802, 1364-1381.
Patterson J. et al.: 2005, *PASP,* 117, I. 837, 1204-1222.
Pearson K. J.: 2006, *MNRAS,* 371(1), 235-244.
Ritter H., Kolb U.: 2005, *VizieR Online Data Catalog,* V/113D.
Skillman D. R., atterson J., Thorstensen J. R.: 1995, *PASP,* 107, 545.
Stefanov S. Y.: 2022, *BAJ,* 36, 21.
Stefanov S. Y., Latev G., Boeva S., Moyseev M.: 2022, *MNRAS,* 516(2), 2775-2781.
Stolz B., Schoembs R.: 1984, *A&A*, 132, 187-194.
Wood M. A., Thomas D. M., Simpson J. C.: 2009, *MNRAS,* 398(4), 2110-2121.

Proceedings of the XIII Bulgarian-Serbian Astronomical Conference
(XIII BSAC) Velingrad, Bulgaria, October 3-7, 2022
Editors: Evgeni Semkov, Milan S. Dimitrijević, Momchil Dechev and Zoran Simić
Publ. Astron. Soc. "Rudjer Bošković" No 25, 2023, 17-25

ON THE ACCRETION EFFICIENCY, DISC'S DENSITY AND TEMPERATURE DISTRIBUTION OF TWO BINARY STARS

DANIELA BONEVA and KRASIMIRA YANKOVA

Space Research and Technology Institute, Bulgarian Academy of Sciences
Acad. Georgi Bonchev Str., Block 1, Sofia 1113, Bulgaria
E-mail: danvasan@space.bas.bg, f7@space.bas.bg

Abstract. In this work, we present our study on the accretion disc' properties of two binary stars: the semi - regular (SR) symbiotic variable NQ Gem and the nova-like system V592 Cas. We apply observational data from both the National Astronomical Observatory (NAO) Rozhen and the American Association of Variable Star Observers (AAVSO). Using color indices from the observations, the color temperatures of the two objects are calculated.

A disc's surface temperature and density distribution in relation to the disc radius are presented. An accretion efficiency is estimated, based on the relation between luminosity and a rate of accretion. We trace changes in efficiency in three possible modes – disc accretion, spherical-like accretion and a two-stream feeding.

We discuss on the currently active mass transfer mechanisms between the binary components. This could be relevant to the type of accretion that is dominant for each of the objects, wind or disc accretion.

1. INTRODUCTION

Studying the properties and parameters of accretion discs could give more detailed view on the whole binary star structure.

The disc's density and surface temperature distributions give an information about the accretion disc's state. To evaluate them, we employ the Kurucz model and the model of D'Alessio and Merin (D'Alessio 1998, 1999). They work with the stars' and discs main parameters, as: mass of the central object M_1; radius of the primary star; disc radius; temperature. The calculations following the model read the assumptions that the heating is both from the stellar radiation and from viscous dissipation due to accretion. The involved into the model equations consider the disc emission. The detailed vertical disc's structure is also included.

An estimation of accretion efficiency is important, because it has a close relation to the emitted energy and has a further role in binaries evolution. An advantage to know the accretion rate is the possibility to estimate the mode of accretion and the possible transition between them.

For the purpose of this paper, we chose to study two binary stars. The first object NQ Gem is classified as a symbiotic star (catalogue of Belczyński et al. (2000)) and belongs to the variables of the Z Andromedae type. They are close binaries that consist of a hot star and a star of a late type. Influenced by the hot star's radiation, an existence of extended envelope is very possible. The Spectral types of those objects are usually C6 or CH3. They are known as "carbon stars" with an excess of carbon in their atmosphere (Bondi 1952).

NQ Gem has a long orbital period and it is found to be 1308 days (Carquillat and Prieur 2008), and its eccentricity is e = 0.182. The mass of the white dwarf or the object's primary component is estimated as M1 ~ 0.6 M⊙ (Luna et al. 2013). This star shows periodicity in its light curves variabilities, but with appearance of irregularity in the brightness with amplitudes 1-2 mag. A long observational period is necessary to detect this peculiarity, usually from 20 to even more than 2000 days. The magnitude range of NQ Gem is in a range 7.4–8.18 in V. Pulsations in its light curves are found, with a pulsation period Ppul ~ 58(±1) days (Gromadzki et al. 2013). The star has also been attached to the group of X-Ray symbiotic binaries (Donahue et al. 1996).

The second object V592 Cas is a Nova like (NL) system and it consists the late-type main-sequence secondary and white dwarf primary. The components are interacting via a warped and tilted accretion disc (Taylor et al. 1998; more generally see also Wood et al. 2000; Murray et al. 2002). V592 Cas is a low-inclination (i =28 deg, Huber et al. 1998) type Cataclysmic variable (CV), with an orbital period of 0.115063(1) d (Taylor et al. 1998), (Witherick et al. 2003) at a distance of 360 pc. The accretion rate of V592 Cas varies in between 9×10^{-9} M⊙ yr^{-1} (Taylor et al. 1998) and 1.3×10^{-8} M⊙ yr^{-1}, estimated by (Ringwald & Naylor 1998). The object exhibits an episodic bipolar wind outflow. V592 Cas shows a behavior of an orbital modulation in the velocity extended UV lines profiles, typical for some cataclysmic variables.

Some systems' parameters of the two objects, necessary for the calculations in this paper, are given in Table 1. The references in brackets are applied below the table.

Table 1: Basic parameters of two binary stars V592 Cas and NQ Gem, used in the calculations: **M_1** and **R_1** are the mass and radius of the primary components in the Solar masses; **L** - the luminosity of V592 Cas, in the solar luminosity; the X-ray luminosity value of NQ Gem is **Lx**; **Teff** is the effective temperature of V592 Cas; $\acute{M}d$ is the disc accretion rate; $\acute{M}sph$ is the spherical accretion rate.

Parameters ------------ **Object**	**M_1 [M⊙]**	$\acute{M}sph$ **[M⊙/y]**	$\acute{M}d$ **[M⊙/y]**	**Porb [days]**	**R_1 [R⊙]**	**L [L⊙]**	**Teff [K]**
NQ Gem	0.6 [6]	---	1.02 x 10^{-8} [6]	1308 [3]	0.03 ≈ Rm [2]	0.2 = Lx [6]	7.2×10^4 ≈ Tm [2]
V 592 Cas	0.75 [4]	5×10^{-10} [5]	1.3×10^{-8} [7]	0.115 [7, 8]	0.01 [4]	0.012 [1]	45000 [4]

[1] Balman, Şölen et al. [2] Boneva and Yankova 2021; [3] Gromadzki et al. 2013; [4] Hoard et al. 2009; [5] Kafka et al. 2009; [6] Luna et al. 2013; [7] Ringwald & Naylor 1998; [8] Witherick et al. 2003.

We consider a shock inflow process, locally onto the accretion stream. In this reason, two additional parameters for NQ Gem are included in the table. **Rm** is the distance from the place with maximum accretion rate to the central object (or primary star). The effective temperature in this place is ***Tm***.

In the next Section 2, consisting of four subsections, we present the obtained results from calculations of density and temperature, color indices and accretion efficiency. We compare the results for V 592 Cas and NQ Gem.

2. RESULTS

2.1. The surface density distribution and temperature variations against the disc radius and time.

The model used for the calculations of the surface density and temperature (see the Introduction) has some limitation of the input parameters' values. One has limit of the primary star effective temperature: Teff = 4000K – 10000K. In this reason it is applicable for NQ Gem, only. The input value of Teff into the model is 4000 K, which is a maximum allowed for this object (Kipper et al. 1996).

In the result, the distribution of both the surface density and two types of the calculated temperatures (the surface Tsurf and the viscous Tvisc) has the form as in Figures 1a and 1b.

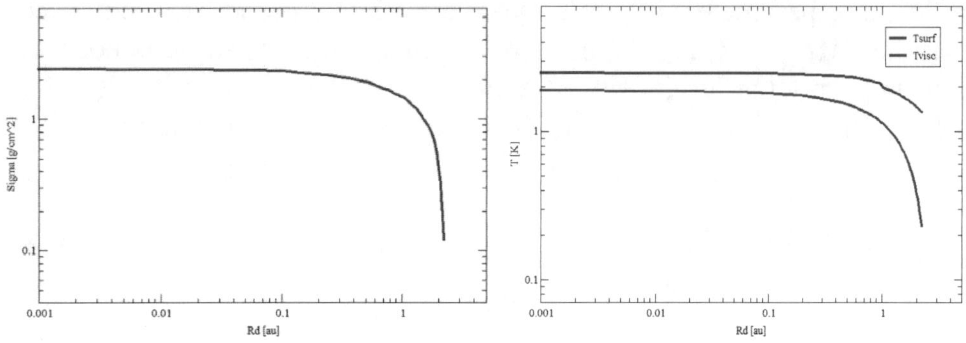

Figure 1: The surface density distribution against maximum disc radius (*left*). The two temperatures (the disc surface Tsurf and the viscous Tvisc) distribution against maximum disc radius (*right*). The input effective temperature for the model calculation is = 4000K. The plotted values in both figures are in a logarithmic scale. The result is applicable for NQ Gem.

As it is seen in the figure, both the surface density and the viscous temperature T_{visc} start to decline at the radii larger that 1 AU, while the surface temperature T_{surf} remains more stable throughout the disc's structure. This could mean that only a small part of the inner disc formation is active for the objects like NQ Gem.

2.2. Color index and color temperature.

To obtain the color index B-V, we use observational data performed with NAO Rozhen for V592 Cas and AAVSO (American Association of Variable Star Observers) for both objects. The observational times are different for two objects, according to their activity periods and available data. For V 592 Cas it is: JD 2459027 – 2459070 and for NQ Gem: JD 2452287-2453480.

The obtained B-V index for NQ Gem is in the range: 1.88 - 2.07(±0.002) and the values are in agreement with the results of Kipper et al. 1996 (2.06) and Zamanov et al. 2022 (2.01).

For V592 Cas, we combined data from both observational sources. In the result, we obtain: (B-V) = 0.15 (± 0.043) - 0.207 (± 0.053). These values differ from the index by Tailor et al. (1998), where the star is estimated as bluer for other observational period. Further, using the B-V index we estimate the color temperature (T_{col}) for two objects, by using the formula of Ballesteros (2012):

$$(1)\ T = 4600K \times \frac{1}{(0.92(B-V)+1.7)} + \frac{1}{(0.92(B-V)+0.62)}$$

Then, all the values of obtained color indices and color temperature are given in Table 2.

Table 2: Color index (B-V) and color temperature Tcol of NQ Gem and V 592 Cas, based on observational data.

Parameter / Object	B-V (min) [mag]	B-V (max) [mag]	Tcol (B-V) [K]	Time [JD]
NQ Gem	2.070 ± 0.005	1.885 ± 0.002	(3098 ± 160) -- (3299 ± 176)	2453310 -- 2453480
V 592 Cas	0.381 ± 0.05	0.151 ± 0.043	(8109 ± 200) -- (8571 ± 200)	2459027 -- 2459070

From the estimated color indices (B-V), we see that the color of both objects are rather red during the studied observational time. With a tendency for higher values of the color temperature, V 592 Cas shows it is a hotter star.

2.3. An accretion efficiency estimation.

The effciency ***η*** expresses the amount of energy gained from the matter with mass ***m***, in units of its mass energy (Kolb 2010). It measures how e□ciently the mass energy, c^2 per unit mass, of the accreted material is converted into radiation. The accretion e□ciency η_{acc} is defined by the expression:

(2) $\eta_{acc} = \frac{GM_1}{R_1 c^2}$,

$$\eta_{acc} = \frac{GM_1}{R_1 c^2} \eta_{acc} = \frac{GM_1}{R_1 c^2}$$

obtained by the expressions of accretion luminosity:

(3) $L_{acc} = \frac{GM_1 \acute{M}}{R_1}$ and $L_{acc} = \eta_{acc} \acute{M} c^2$

Where M_1 and R_1 are the mass and radius of the primary star; G is the gravitational constant; c is the speed of light; $\acute{M}$ is the accretion rate. We also use the relation between the luminosity L and the effective temperature T_{eff}, by the Stefan-Boltzmann's law:

(4) $L = 4\pi R_{R1}^2 \sigma T_{eff}$

The changes in efficiency could be traced in three possible modes: disc accretion - ***η***, spherical-like accretion - *ξ* and a two-stream feeding - *ξ*+***η***. Since, equation (2) gives the efficiency of the disc' accretion, we should obtain expressions for the other two modes. First, we apply equations of the related accretion rates: disc and spherical. If we denote the accretion rate in general case with $\acute{M}$, then $\acute{M}_d$ expresses the disc accretion, by the equation (5), based on (Frank et al 2002):

$$(5)\ \acute{M}_d = 2\pi R_{R1}\Sigma(-V_r),$$

where *Vr* is the radial velocity of the stream, Σ is the surface density; R_{R1} represents the distance to the central object (in some cases, it coincides with the radius of the primary or the disc inner radius). Further, the spherical accretion rate could be expressed as (Bondi 1952, Edgar 2004):

$$(6)\ \acute{M}_{sph} = \pi R_{R1}^2 \rho(-V_r),$$

ρ is a volume density in this case. The ratio of the spherical accretion efficiency ***ξ*** to the efficiency of the disc accretion - ***η*** can be expressed as a relation between the two types of accretion rate:

$$(7)\ \frac{\xi}{\eta} = \frac{\acute{M}_d}{\acute{M}_{sph}}$$

Then for the efficiency of the spherical accretion we can write:

$$(8)\ \xi \approx \left(\frac{4H_d}{R_d}\right)\eta$$

Where ***H_d*** is the full half disc thickness and the whole disc radius is ***R_d***. We obtain the minimum and maximum values of the ratio, as: $4H_d/R_d = [0.04; 0.2]$. Our estimation for NQ Gem gives the value ≈ 0.16 and for V 592 Cas it is ≈ 0.14.

Next, the parameters' values of V 592 Cas and NQ Gem (see Table 1) are applied into the Equations (2-8). An important note for the calculations of NQ Gem is that we are using the effective temperature ***Tm***, measured in the place with maximum accretion rate at ***Rm*** (see the explanations in the Introduction).

The obtained values of the accretion efficiency in the three suggested accretion modes are given in Table 3:

Table 3: The mean values of accretion efficiency calculated for both objects in the three suggested accretion modes: ξ for spherical accretion, η - for disc accretion, and a sum of both η + ξ expresses the two-stream feed accretion.

Object/Mode	**1**	**2**	**3**
X 10 $^{-4}$	ξ	η	ξ+η
NQ Gem	0.20 ± 0.02	1.28 ± 0.12	1.50 ± 0.14
V 592 Cas	0.16 ± 0.02	1.17 ± 0.11	1.28 ± 0.13

We see that the values for NQ Gem and V 592 Cas are similar. The distribution of accretion efficiency throughout the development of accretion modes is represented in Figure (2).

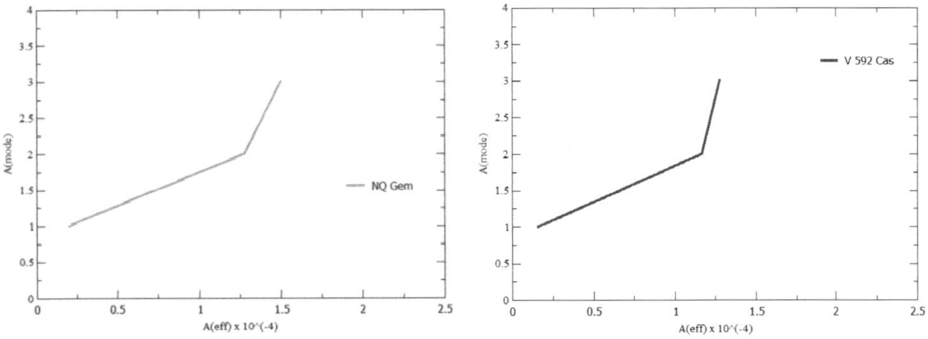

Figure 2: A growth of accretion efficiency A(eff) against the development of accretion modes A(mode), 1(ξ), 2(η) and 3(η + ξ), for two objects. For NQ Gem (*left*) and for V 592 Cas (*right*).

These two empirically created figures depict a slow and smoothly increase of the spherical accretion efficiency, while the growth of the efficiency in the disc accretion mode is sharper.

2.4. Mass-transfer between the components.

The matter transfers between the binary stars' components usually in two currently known ways. When one of the stars increases in the radius and fills its Roche lobe, the matter overflows through the Lagrange point L1 (Frank et al. 2002). The further product of this interaction could be the formation of accretion flow or an accretion disc. In some binary stars, one of the components throws out much of its material in a form of a stellar wind, which could be later attracted by the gravitational field of the primary star. By a relation to the accretion efficiency parameters' values, we could give a point to a currently active mass-transfer mechanism in the two binary stars studied here.

In the Novae like object V 592 Cas, a precessing accretion disc around the primary component is observed (Long 2003). This might be not a main, but a possible reason for the relatively low values of the disc's accretion efficiency. We give a suggestion that in this binary, the stream from the secondary is transferring through the Lagrange point L1. A pseudo-spherical inflow is also possible to contact the outer parts of the accretion disc.

A reason of the relatively higher efficiency of NQ Gem could be that the parameters' calculations are performed for the zone with a maximum accretion rate. The activity of NQ Gem in X-ray luminosity could be contributed by the mass transfer via the stellar wind (see also Boneva & Yankova 2021).

3. CONCLUSIONS

In this paper, we presented some accretion disc properties for two binary stars: V 592 Cas and NQ Gem. Based on the irradiated disc model, the density and temperature distribution is estimated for NQ Gem.

Using observational data, we compared the color indices and color temperatures of V 592 Cas and NQ Gem.

We estimated the accretion efficiency variations of the two studied objects, on the three considered accretion modes: spherical accretion, disc accretion and a two-stream feed accretion. The results showed that the growth of disc efficiency of V592 Cas is sharper than that of NQ Gem, but both objects have a smoothly development of accretion efficiency throughout the modes.

Acknowledgements

This work is partially supported by the grant "Binary stars with compact object", КП-06-Н28/2 08.12.2018 (Bulgarian National Science Fund).

The authors thank the AAVSO (American Association of Variable Star Observers) for providing the data of Light Curve Generator, contributed by observers worldwide and used in this research.

References

Ballesteros F. J.: 2012, *EPL*, 97, 34008.
Balman Ş., Patrick G., Sion E. M.: 2014, *ApJ*, 794:84, pp 12
Belczyński K., Mikołajewska J., Munari U., Ivison R. J., Friedjung M.: 2000, *A&AS*, 146, 407
Bondi H. 1952, *MNRAS*, 112, 195.
Carquillat J. M., Prieur J. L.: 2008, *AN*, 329, 44.
D'Alessio P., Cant'o J., Calvet N., Lizano S.: 1998, *ApJ*, 500, 411
D'Alessio P., Calvet N., Hartmann L., Lizano S., Cant´o J.: 1999, *ApJ*, 527, 893
Donahue R. A., Saar S. H., Baliunas S. L.: 1996, *ApJ*, 466,384.
Edgar R.: 2004, *New Astronomy Reviews*, 48, 843.
Frank J., King A., Raine D. J.: 2002, *Accretion Power in Astrophysics:3rd Edition.*
Gromadzki M., Mikołajewska J., Soszyński I.: 2013, *AcA*, 63, 405.
Hoard D. W., Kafka S., Wachter S., et al.: 2009, *ApJ*, 693, 236
Huber M. E., Howell S. B., Ciardi D. R., Fried R.: 1998, *PASP*, 110, 784
Kafka S., Hoard D. W., Honeycutt R. K., Deliyannis C. P.: 2009, *AJ*, v 137, 1, pp. 197-206.
Kipper T., Jorgensen U.G., Klochkova V.G., Panchuk V.E.: 1996, *A&A,* 306, p.489
Kolb U.: 2010, *Extreme Environment Astrophysics,* Cambridge: Cambridge University Press.
Long K.: 2003, *FUSE Proposal* ID D11, 2003fuse.prop.D114L
Luna G. J. M., Sokoloski J. L., Mukai K., Nelson T.: 2013, *A&A*, 559, p.6
Murray J. R., Chakrabarty D., Wynn G., Kramer L.: 2002, *MNRAS*, 335, 247
Ringwald F. A., Naylor T.: 1998, *AJ*, 115, 286

Taylor C. J., Thorstensen J. R., Patterson J., et al.: 1998, *PASP*, 110, 1148
Witherick D. K., Prinja R. K., Howell S. B., Wagner R. M.: 2003, *MNRAS*, 346, 861, W03
Wood M. A., Montgomery M. M., Simpson J. C.: 2000, *ApJ*, 535, L39
Zamanov R. K., Bode M. F., Melo C. H. F., Bachev R., Gomboc A., Porter J., Pritchard J.: 2007, *MNRAS*, 380, 1053

Proceedings of the XIII Bulgarian-Serbian Astronomical Conference
(XIII BSAC) Velingrad, Bulgaria, October 3-7, 2022
Editors: Evgeni Semkov, Milan S. Dimitrijević, Momchil Dechev and Zoran Simić
Publ. Astron. Soc. "Rudjer Bošković" No 25, 2023, 27-34

ANALYSIS OF CALCULATED STARK BROADENING PARAMETERS OF SINGLY IONIZED SILICON LINES

MAGDALENA D. CHRISTOVA[1], MILAN S. DIMITRIJEVIĆ[2,3] and SYLVIE SAHAL-BRÉCHOT[3]

[1]*Department of Applied Physics, FAMI, Technical University of Sofia, blvd. Kl. Ohridski 10, 1000 Sofia, Bulgaria*
[2]*Astronomical Observatory, Volgina 7, 11060 Belgrade, Serbia*
[3]*LERMA, Observatoire de Paris, Université PSL, CNRS, Sorbonne Université, 5 Place Jules Janssen, 92190 Meudon, France*
E-mail: mchristo@tu-sofia.bg, mdimitrijevic@aob.bg.ac.rs, sylvie.sahal-brechot@obspm.fr

Abstract. Analysis of calculated Stark broadening parameters (full width at half maximum and shift) of Si II spectral lines is presented. The importance of different interactions between emiters and perturbers (electrons, protons, and helium ions) is discussed. The conditions of interest cover temperature values 5 000 K, 10 000 K, 20 000 K, 30 000 K, 50 000 K, 100 000 K and perturber density from 10^{14} cm^{-3} to 10^{20} cm^{-3}. The results could be useful for astrophysical purposes, laboratory plasma and inertial fusion research, as well as in industrial plasma technologies.

1. INTRODUCTION

The considerable interest to the Stark broadening of Si II spectral lines arises from the large cosmic abundance of silicon, which is at the sixth place in the stellar abundance distribution of elements (SAD) (Unsöld and Bashek, 1967). The importance of some strong ion lines for the spectroscopy of solar atmosphere is outlined in Hey (1977). Singly ionized silicon lines are present in hot stars from B and A type, and also in O type stars and white dwarfs (Peytremann, 1972). Observations of Si II and Si III lines in the spectrum of peculiar stars, particularly in silicon Ap-stars are reported in (Dimitrijević, 1983 and references therein). The evaluation of the physical conditions in stellar plasma needs reliable Stark broadening data. As it is reported in Lanz et al. (1988) the silicon spectrum is dominant for stars from A0 to B3 spectral type with effective temperatures from 10 000 K to 20 000 K. Despite the resonance multiplets, strong lines are observed in the visible and ultraviolet diapason. These lines were used for silicon abundance determination in Mihalas and Henshow (1966) and Megessier (1971). Kamp (1978) and Lennon et al. (1986) investigated silicon absorption in hot stars using these strong multiplets. According to Vince et al. (1985), Stark broadening plays a role in the wings of spectral lines in solar spectrum. With the development of the instruments, reliable data are obtained for faint and less intensive lines, corresponding to higher transitions which are sensitive to the plasma conditions. As Seaton mention in his work from 1987, Stark broadening results contribute to the development of physics of stellar interiors where energy transport models require knowledge of relevant atomic process and radiative opacities.

In laboratory plasmas, silicon is a principal contaminant and silicon lines appear in the radiation spectrum (Hey, 1978; Dimitrijević, 1983). Thus, it is useful to perform spectroscopic study of silicon lines for plasma diagnostics. There are many publications on the Stark broadening measurements (Konjević et al., 1970; Purić et al., 1974; Lesage and Miller, 1975; Lesage et al., 1977; Chiang and Griem 1978; Lesage et al., 1983; Kusch and Schröder, 1982; Peréz et al., 1993; Wollschläger et al., 1997; Gonzalez et al., 2002; Lesage and Redon, 2004, Bukvić et al., 2009; Gavanski et al., 2016). It should be noted that a large scatter of experimental results is found (Konjević and Wiese, 1990). Recently, a comprehensive review on the application of silicon laser induced breakdown spectroscopy (LIBS) is published in Ivković et al. (2017). There are also many papers that deal with calculation of Stark broadening parameters of singly ionized silicon lines (Griem, 1974, Hey, 1978, Dimitrijević, 1983, Lanz et al., 1988, Gavanski et al., 2016). In Dimitrijević et al. (2023), we calculate Stark widths of 13 Si II multiplets for temperatures of 5000 K, 10 000 K, 20 000 K, 40 000 K and 80 000 K, and for perturber density of 10^{17} cm^{-3} applying modified semiempirical method (MSE) described in Dimitrijević et al. (1980, 1986). Comparison between measured Stark broadening parameters (width and shift) and our calculated results for Si II lines ($3s^24s\ ^2S_{1/2} - 3s^24p\ ^2P^o_{3/2}$, $\lambda = 6356.9$ Å; $3s^24p\ ^2P^o_{3/2} - 3s^25s\ ^2S_{1/2}$, $\lambda = 5973.4$ Å; $3s^24p\ ^2P^o_{3/2} - 3s^24d\ ^2D_{5/2}$, $\lambda = 5052.4$ Å and $3s^23d\ ^2D_{5/2} - 3s^24d\ ^2F^o_{7/2}$, $\lambda = 4130.9$ Å) is given in Dimitrijević et al. (2022).

2. THEORY

Stark broadening is a type of pressure broadening of spectral lines emitted or absorbed in plasma environment. During the process of emission (absorption) of a photon the atom (ion) interacts with surrounding charged particles as electrons, protons, and ions which broaden and shift the spectral line. We apply semiclassical perturbation theory (Sahal-Bréchot, 1969a; Sahal-Bréchot, 1969b) and further improvements (Sahal-Bréchot, 1974; Fleurier et al., 1977; Dimtrijević et al., 1991; Sahal-Bréchot, 1991; Sahal-Bréchot, 2014). The semiclassical Sahal-Bréchot theory is in the collisional approach where the atom (ion) is examined as quantum system and the charged particles as classical ones. The Stark broadening parameters (full width at half-maximum of intensity and shift of the line) are expressed by the following formulae:

$$W = 2n_e \int_0^\infty \mathrm{v}f(v)\,\mathrm{dv}\left[\sum_{i' \neq i} \sigma_{ii'}(v) + \sum_{f' \neq f} \sigma_{ff'}(v) + \sigma_{el}\right]$$

$$d = \int_0^\infty \mathrm{v}f(v)\,\mathrm{dv} \int_{R_3}^{R_d} 2\pi\rho d\rho \sin 2\phi_p$$

where i and f denote the initial and final level of the studied transition; i' and f' are perturbing levels; n_e electron density; v velocity of the corresponding charged particle, and $f(v)$ is the Maxwellian distribution of electron velocities. The cross sections $\sigma_{ii'}(v)$ concern inelastic collisions between atoms (ions) in initial energy level and perturbers. The cross section $\sigma_{ff'}(v)$ gives inelastic collisions for transition between f and f'' atomic energy levels. These terms could be presented by an integration of the transition probability $P_{ii'}$ (respectively $P_{ff'}$), over the impact parameter ρ:

$$\sum_{i' \neq i} \sigma_{ii'}(v) = \frac{1}{2}\pi R_1^2 + \int_{R_1}^{R_d} 2\pi\rho d\rho \sum_{i' \neq i} P_{ii'}(\rho, v)$$

The contribution of elastic collisions between atoms (ions) and charged particles could be estimated by the expressions:

$$\sigma_{el} = 2\pi R_2^2 + \int_{R_2}^{R_d} 8\pi\rho d\rho \sin^2\delta + \sigma_r$$

$$\delta = \left(\phi_p^2 + \phi_q^2\right)^{1/2}$$

In the above expression δ gives the phase shift which components φ_p and φ_q, describe atom (ion)-perturber elastic collisions via polarization and quadrupole potentials, respectively. Additional clarifications on the symmetrization and cut-off parameters procedures (R_1, R_2, R_3, the Debye cut-off R_d) are available in Sahal-Bréchot, (1969b). The cross section σ_r treats Feshbach resonances (Fleurier et al., 1977; Sahal-Bréchot, 2021). To be useful for both, astrophysical purposes and laboratory spectroscopic diagnostics, our calculations for broadening parameters of Si II lines include collisions of ions with electrons, protons, and ionized helium. The proposed work includes analysis of the perturber-impact component in Stark width and shift of singly ionized silicon lines and behavior of full width and shift within a spectral series. The plasma conditions of interest are consistent with astronomical observations and laboratory measurements.

3. RESULTS

In this study we are interested in: 1) contribution of different type of collisions to the Stark broadening in the whole temperature interval; 2) behavior of width and shift within a spectral series. One of the examined spectral lines corresponds to transition $3s^23d\ ^2D_{5/2} - 3s^24d\ ^2F^o_{7/2}$, λ = 4130.9 Å. The investigated spectral series is $3s^23p - 3s^2ns$ for n = 4 - 8. Temperature interval includes following values: 5 000 K, 10 000 K, 20 000 K, 30 000 K, 50 000 K, and 100 000 K. Perturber density varies from 10^{14} cm^{-3} to 10^{20} cm^{-3}.

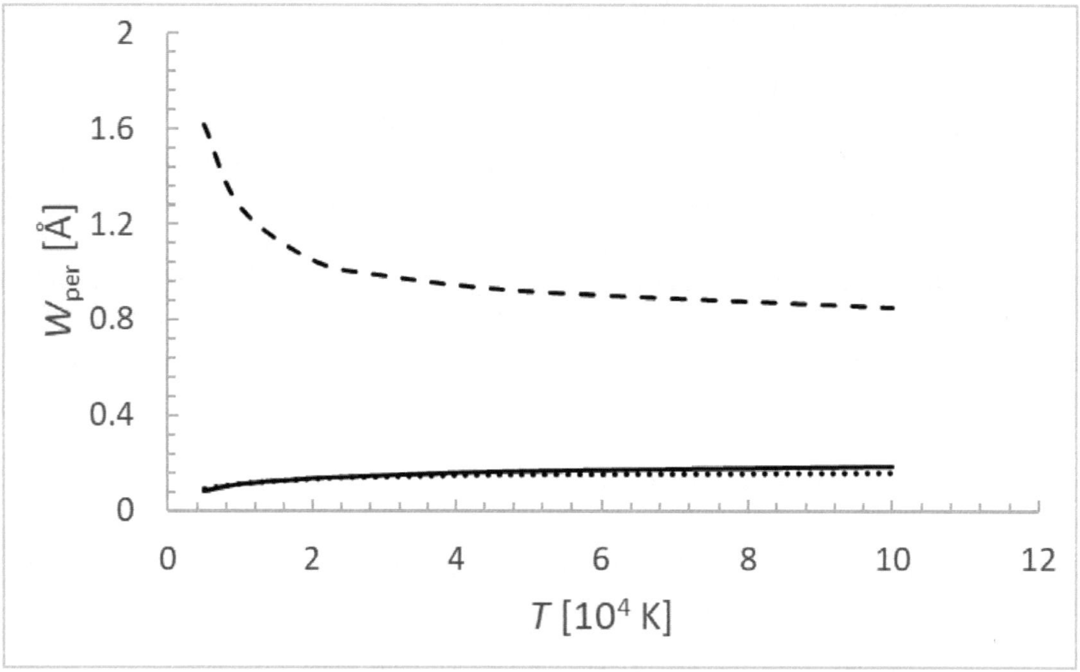

Figure 1: Behavior of electron-impact (dashed line), proton-impact (solid line) and helium ion-impact (dot line) width of Si II $3s^23d\ ^2D_{5/2} - 3s^24d\ ^2F^o_{7/2}$ spectral line (λ = 4130.9 Å) in whole temperature interval. Perturber density is 10^{17} cm^{-3}.

An example for the perturber-impact width and shift behavior versus temperature is the behavior of Stark width and shift for λ = 4130.9 Å, illustrated in figures 1 and 2. Electron-impact width is considerable for all temperatures and significantly bigger than other two width components even that it decreases with T. It's decreasing is prominent for lower temperatures, while after 20 000 K this component decreases slowly. Electron-impact width forms full width at half-maximum in whole temperature interval. Both, proton-impact, and helium ion-impact components almost coincide. They increase slowly for higher temperatures. For small temperatures interactions with helium ionized atoms are a little bit more effective, while for bigger T-values they are less effective than interactions with protons.

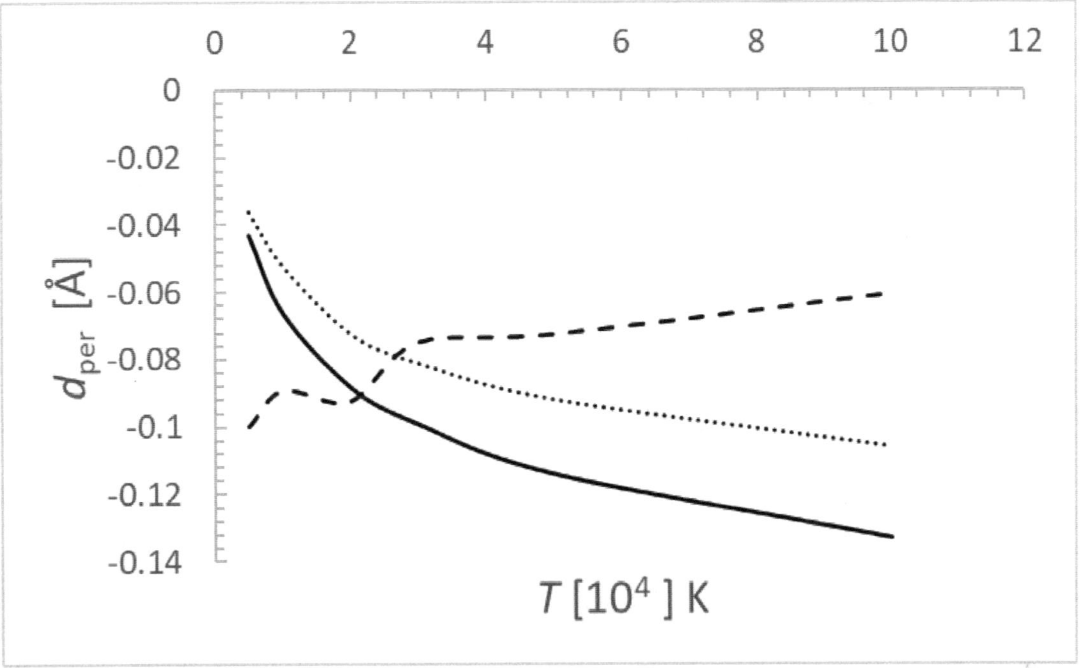

Figure 2: Behavior of electron-impact (dashed line), proton-impact (solid line) and helium ion-impact (dot line) shift of Si II $3s^23d\ ^2D_{5/2} - 3s^24d\ ^2F^o_{7/2}$ spectral line ($\lambda = 4130.9$ Å) in whole temperature interval. Perturber density is 10^{17} cm^{-3}.

Stark shift of this line is negative (blue shifted). Completely different picture is obtained for their shift components. Up to 20 000 K electron-impact shift predominates. Some oscillations of this component are observable. Around 20 000 K all three components are comparable. From 20 000 K to 100 000 K other two components, proton-impact and helium ion-impact are larger. They increase with *T* and follow almost same trend. Proton width component is bigger than ionized helium one for all *T*-values as the difference increases. The results show that the line is broadened by impacts with electrons and shifted due to interactions with all type of perturbers.

Knowledge of Stark broadening parameters for one spectral series is very important. Via interpolation or extrapolation of available results it ensures estimation of width and shift for lines which are difficult to observe, to measure or to calculate their parameters due to absence of atomic data. In the next two figures we present electron-impact width and shift for spectral lines belonging to Si II $3s^23p - 3s^2ns$ series for principal quantum number $n = 4 - 8$. For lines originated from higher energy levels ($n = 7, 8$) collisional approach is no longer applicable for protons and helium ions interactions with singly ionized silicon ions. It is the same for $3s^23p - 3s^26s$ transition concerning temperature up to 30 000 K. There is a notable increasing of electron impact width for $n > 6$. All shift values are positive, and the same significant increasing of electron-impact shift is demonstrated for $n > 6$.

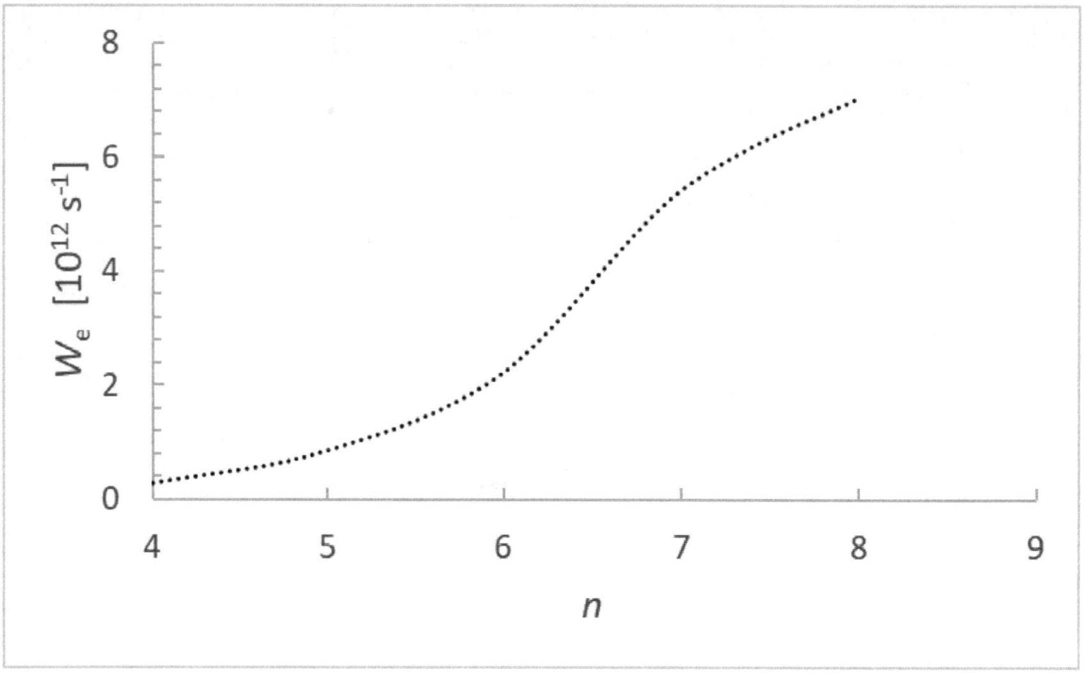

Figure 3: Electron-impact width versus principal quantum number within one spectral series $3s^23p$ – $3s^2ns$ for n = 4 – 8. The temperature is 20 000 K and the perturber density - 10^{17} cm^{-3}.

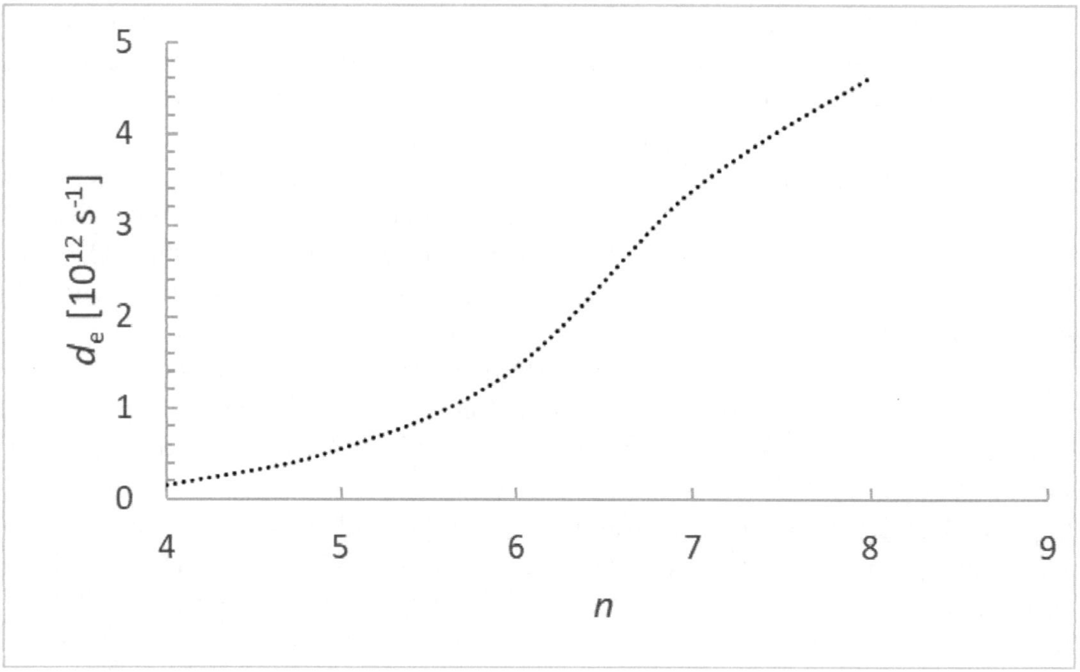

Figure 4: Electron-impact shift versus principal quantum number within one spectral series $3s^23p$ – $3s^2ns$ for n = 4 – 8. The temperature is 20 000 K and the perturber density - 10^{17} cm^{-3}.

4. CONCLUSIONS

It has been demonstrated in this work the behavior with temperature of Stark widths and shifts of Si II spectral lines, due to collisions with electrons, protons and singly charged helium ions. Also, Si II electron-impact widths and shifts behavior within a spectral series has been investigated.

Acknowledgments

This work has been supported with a STSM visit grant CA16214-48065 for M.S.D. within the framework of COST Action CA 18104 "Revealing the Milky Way with GAIA".

Partial support from Faculty of Applied Mathematics and Informatics, Technical University of Sofia, Bulgaria is also acknowledged. This work has been partially supported and by the Paris Observatory, the CNRS, and the PNPS (Programme National de Physique Stellaire, INSU-CNRS), France.

References

Bukvić S., Djeniže S., Srećković A.: 2009, *A&A*, 508, 491.
Chiang T., Griem H. R.: 1978, *Phys. Rev. A*, 18, 1169.
Dimitrijević M. S.: 1983, *A&A*, 127, 68.
Dimitrijević M. S., Christova M. D., Sahal-Bréchot S.: 2022, *Contrib. Astron. Obs. Scalnaté Pleso*, 52/3, 25.
Dimitrijević M. S., Christova M. D., Yubero C.: 2023, *Adv. Sp. Res.*, 71, 1275.
Dimitrijević M. S., Konjević N.: 1980, *J. Quant. Spectrosc. Radiat. Transfer*, 24, 451.
Dimitrijević M. S., Kršljanin, V.: 1986, *A&A*, 165, 269.
Dimitrijević M. S., Sahal-Bréchot S.: 2014, *Atoms*, 2, 357.
Dimitrijević M. S., Sahal-Bréchot S., Bommier V.: 1991, *A&AS*, 89, 581.
Fleurier C., Sahal-Bréchot S., Chapelle J.: 1977, *J. Quant. Spectrosc. Radiat. Transfer*, 17, 595.
Gavanski L., Belmonte M. T., Savić I., Djurović S.: 2016, *MNRAS*, 457, 4038.
Gonzalez V. R., Aparicio J. A., del-Val A., Mar S.: 2002, *J. Phys. B: At. Mol. Opt. Phys.*, 35, 3557.
Griem H. R.: 1974, *Spectral line Broadening by Plasmas,* McGraw-Hill, New York.
Hey J. D.: 1978, *J. Quant. Spectrosc. Radiat. Transfer*, 18, 425.
Ivković M., Konjević N.: 2017, *Spectrochem. Acta Part B*, 131, 79.
Lanz T., Dimitrijević M. S., Artru M. C.: 1988, *A&A*, 192, 249.
Lesage A., Sahal-Bréchot S., Miller M. H.: 1977, *Phys. Rev. A*, 16, 1617.
Lesage A., Rathore B. A., Lakićević I. S., Purić J.: 1983, *Phys. Rev. A*, 28, 2264.
Lesage A., Redon R.: 2004, *A&A*, 418, 765.
Kamp L. W.: 1978, *Astrophys. J. Suppl. Ser.*, 36, 143.
Konjević N., Purić J., Ćirković Lj., Labat J.: 1970, *J. Phys. B*, 3, 999.
Konjević N., Wiese W. L.: 1990, *J. Phys. Chem. Ref. Data*, 19, 1307.
Kusch H. J., Schröder K.: 1982, *A&A*, 116, 255.
Lennon D. J., Lynas-Gray A. E., Brown P. J. F., Dufton P. L.: 1986, *MNRAS*, 222, 719.

Lesage A., Miller M.: 1975, *Academie des Sciences Paris Comptes Rendus Serie B Sciences Physiques*, 280, 645
Mégessier C.: 1971, *A&A*, 10, 332.
Mihalas D., Henshaw J. L.: 1966, *Astrophys. J.*, 144, 25.
Peréz C., de la Rosa I., de Frutos A. M., Mar S.: 1993, *Phys. Rev. E*, 47, 756.
Peytremann E.: 1972, *A&A*, 17, 76.
Purić J., Djeniže S., Labat J., Ćirković, Lj.: 1974, *Z. Phys.*, 267, 71.
Sahal-Bréchot S.: 1969a, *A&A*, 1, 91.
Sahal-Bréchot S.: 1969b, *A&A*, 2, 322.
Sahal-Bréchot S.: 1974, *A&A,* 35, 319.
Sahal-Bréchot S.: 1991, *A&A*, 245, 322.
Sahal-Bréchot S.: 2021, *Atoms*, 9, 29.
Sahal-Bréchot S., Dimitrijević M. S., Ben Nessib N.: 2014, *Atoms*, 2, 225.
Seaton M. J.: 1987, *J. Phys. B: At. Mol. Phys.,* 20, 6363.
Unsöld A., Bashek B.: 1967, *Der Neue Kosmos,* Springer, Berlin, Göttingen, Heidelberg.
Vince I., Dimitrijević M. S., Kršljanin V.: 1985, *Progress in Stellar Spectral Line Formation Theory*, eds. Beckmann et al., 373.
Wollschläger F., Mitsching J., Meiners D., Depiesse M., Richou J., Lesage A.: 1997, *J. Quant. Spectrosc. Radiat. Transfer*, 58, 135.

Proceedings of the XIII Bulgarian-Serbian Astronomical Conference
(XIII BSAC) Velingrad, Bulgaria, October 3-7, 2022
Editors: Evgeni Semkov, Milan S. Dimitrijević, Momchil Dechev and Zoran Simić
Publ. Astron. Soc. "Rudjer Bošković" No 25, 2023, 35-42

BULGARIAN-SERBIAN COOPERATION: CCD OBSERVATIONS OF VISUAL DOUBLE STARS IN THE PERIOD 2004-2022

ZORICA CVETKOVIĆ[1], RADE PAVLOVIĆ[1] and
SVETLANA BOEVA[2]

[1]*Astronomical Observatory, Volgina 7, 11060 Belgrade, Serbia*
[2]*Institute of Mathematics and Informatics, Bulgarian Academy of Sciences*
Acad. Georgi Bonchev Str., Block 8, Sofia 1113, Bulgaria
E-mail: zorica@aob.rs, rpavlovic@aob.rs, sboeva@astro.bas.bg

Abstract. Visual binaries have been studied for more than 200 years, from wide pairs resolvable by the naked eye toward more close ones being resolved by using various observational techniques of high angular resolution such as speckle interferometry, lunar occultations, adaptive optics, etc. With classic CCD technique it is not possible to resolve systems with small separation between the components. Wide stellar systems, with a separation greater than 2 arcseconds, have large orbital periods in general. For many of them there are no many observations, the observations over a short orbital arc or they have a low accuracy. Because of this, as many as possible new observations of wide double stars should be performed.

The Serbian-Bulgarian cooperation concerning the study of visual double and multiple stars started in 2004 with CCD observations of such objects. Here we present a brief overview of the most interesting results obtained during our collaboration. Begun as a regional Balkan project, this cooperation in 2012 was continued in the form of a bilateral cooperation between the Bulgarian Academy of Sciences and Serbian Academy of Sciences and Arts.

1. INTRODUCTION

Systematic observations of double and multiple stars have been carried out for more than 200 years. Their study makes possible more accurate determination of masses and distances, as well as a better understanding of stellar formation and evolution. The results of measurements have been collected and the corresponding database is kept by the United States Naval Observatory. The Washington Double Star Catalogue (WDS)[1] has been constantly updated and contains the data for

[1] http://www.astro.gsu.edu/wds/

more than 154,000 pairs. The orbital elements have been obtained for a small number of them, about 3450, and they can be found in the Sixth Catalog of Orbits of Visual Binary Stars, (ORB6)[2]. For about 1280 pairs it has been established that their components are not at the same distance, i.e., that they are not gravitationally bound. For them the linear elements of motion have been determined and they are given in the Second Catalog of Rectilinear Elements (LIN2)[3].

Observations of double and multiple stars have been carried out at the Astronomical Observatory of Belgrade for more than 70 years. Until 1994, micrometer measurements of relative coordinates between component pairs were made, and after that observations were made with CCD cameras. Since 2004 till now a group of astronomers from Belgrade Observatory have visited many times the National Astronomical Observatory Rozhen (NAOR) in Bulgaria and taken frames of visual double and multiple stars. Series of observations have been made with a CCD camera attached to the 2-m telescope. Only during the observations made in 2004 the CCD camera Photometrics CE200A was used. Later observations made between 2005 and 2018 have been performed with the CCD camera VersArray 1300B. For measurements made since 2019, a CCD camera Andor iKon-L was used. Until now we have made 27 series of observations of the above mentioned stars. Table 1 shows the dates of observation series at NAO Rozhen, the number of observation nights, the number of observation pairs and the references in which the results were published. A total of 4398 pairs were measured (during the first 24 series of observations). The results: position angles, angular separations, orbital and linear solutions, have been published in Pavlović et al. (2005, 2013), Cvetković et al. (2006, 2007, 2010, 2011, 2015, 2016, 2017, 2018, 2019, 2021). The measurements of the frames for 137 pairs obtained in 2021 and in 2022 (the last three series) to be published soon.

In June of 2011 the first CCD observations of double and multiple stars with the 60 cm telescope from the new Astronomical Station on the mountain of Vidojevica (ASV) were started. During the autumn of 2011 we observed the same objects at both NAOR and ASV. We noticed that the measured separations (ρ_{NAOR}, ρ_{ASV}) differ for the same pairs of stars and the differences increase with increasing angular separation. Therefore, we measured the angular separations between the images of stars visible in our CCD frames. The separation depends on the pixel scale, i.e. the focal length of the telescope. The result of determining the focal length of the 60 cm telescope at the ASV more precisely is given in the paper Cvetković et al. (2012b), and for the 2 m NAOR telescope in the paper Cvetković et al. (2013).

2 http://www.astro.gsu.edu/wds/orb6/orb6frame.html
3 http://www.astro.gsu.edu/wds/lin2/lelements.html

Table 1: An overview of the number of observation nights, the number of observation pairs and the references in which the results were published

Date	Numb. Observ. Nights	Number of pairs	References
October 2004	1	27	Serb. Astron. J. **171**, 2005
October 2005	1.5	35	Serb. Astron. J. **172**, 2006
December 2006	1	67	Serb. Astron. J. **174**, 2007
July 2009	1	35	Serb. Astron. J. **180**, 2010
September 2010	3	202	Astron. Journal **142**, 2011
March and October 2011	3	337	Astron. Journal **146**, 2013
April and November 2012	6	453	Astron. Journal **149**, 2015
April and October 2013 March and October 2014	7	721	Astron. Journal **151**, 2016
April 2015	4	801	Astron. Journal **153**, 2017
March/April and October 2016	4	450	Astron. Journal **156**, 2018
April and October 2017 April and November 2018	5	939	Astron. Journal **158**, 2019
March and November 2019 March and November 2020	4	331	Astron. Journal **162**, 2021
		4398	
April and October 2021	2	74	unpublished
March 2022	3	63	unpublished

2. RESULTS

Using only our observations we analyzed a multiple system registered in ADS - Aitken Double Stars catalogue as ADS 48. Its number in the Washington Double Star Catalog is 00057+4549. Our aim is to establish which of the seven components (A, B, C, D, E, P and F) are gravitationally bound, i.e. have an orbital motion around the mass center, and which of them are mutually very distant in space so that only their projections are close in the field of view. We used the measuring results from our CCD frames obtained between 1994 and 2011. The first CCD frames of ADS 48 multiple system at our disposal were obtained in 1994. We also obtained frames of this system at NAOR in 2004, 2005, 2006, 2010 and 2011, as well as three times at ASV in 2011. The selected CCD frames obtained from 1994 to 2011 as well as the detailed analysis of the system ADS 48 is given in the paper Cvetković et al. (2012a). The conclusions combined with the criteria based on celestial mechanics lead us to the following: i) within the system ADS 48 only stars A and B are gravitationally bound; ii) component F has common proper motion with A and B, but is not bound to them; iii) all other components considered here form optical pairs with AB.

Using the measurements obtained by us from the CCD observations at NAO Rozhen we have calculated the first orbits for two binaries: WDS 00152+2722 = J 868 given in Novaković (2007) and WDS 07106+1543 = J 703 given in Cvetković and Ninković (2008).

Three orbits were corrected for pairs WDS 03342+4837 = BU 787 AB, WDS 21289+1105 = STF 2799 AB and WDS 22234+3228 = WOR 11. The orbital elements are given in Cvetković et al. (2011).

We analyzed one more pair, WDS 03342+4837 = BU 787 AB. This pair belongs to a multiple system. It was discovered in 1881 and denoted as BU 787 AB. The components D and E were discovered later. The first orbit was calculated (Erceg 1984). The orbital elements and the new Hipparcos parallax yield $2600M_\odot$ for the total mass of the system. This is an unrealistic value. Four measurements were performed at NAOR enable us to recalculate the orbit. But, all measurements show a linear trend and we are the first who calculated the linear solution for this pair. Additional analysis indicated that it is most likely an optical pair. Both recalculated orbit and linear solution are published in Cvetković et al. (2011).

Preliminary orbital elements for the pair J 703 published in 2008 were the first calculation of the orbital elements for it. There are three measurements for this pair obtained after 2008: two in 2011 and one in 2012. They have significant residuals in separation from the orbit. Also, the measurements show a linear trend and we are the first who calculated the linear solution for this pair. It is given in Cvetković et al. (2015).

In addition to these two pairs, BU 787 AB and J 703, measurements for another 30 observed pairs show a linear trend and we are the first who calculated their linear solutions. The results of their analysis are given in Table 2 and in the next section.

3. THE NATURE OF MOVEMENTS

Two linear elements, V_x and V_y , are used to calculate the velocity V of relative motion of the secondary with respect to the primary. Also, we can calculate the relative proper motion μ_{rel} for 32 pairs by using the components of the proper motion, $\mu_{\alpha\cos\delta}$ and μ_δ, for primary A and secondary B. They are taken from Gaia DR3[4]. The Gaia DR3 catalogue is available from June 2022. It contains parallaxes for both components of all pairs from our sample. Table 2 gives data for the 32 pairs with linear solutions. The columns are: 1 − WDS designation; 2 − discovery designation on whose basis it is possible to identify each pair; 3 − the relative proper motion μ_{rel},; 4 − the velocity V of relative motion of the secondary with respect to the primary; 5 and 6 − the primary parallax π_A and the secondary parallax π_B, respectively, from Gaia DR3.

[4] http://vizier.u-strasbg.fr/viz-bin/VizieR-3?-source=I/355/gaiadr3

Table 2: The relative proper motion μ_{rel}, the velocity V of relative motion of the secondary with respect to the primary, the primary parallax π_A and the secondary parallax π_B for 32 pairs with their WDS numbers and discoverer's designations

WDS Design.	Discoverer	μ_{rel}	V	π_A	π_B
α,δ (2000)	Designation	("/yr)	("/yr)	(mas)	(mas)
00057+4549	STT 547AC	0.8984	0.8991	86.800	0.923
00057+4549	STT 547AD	0.8996	0.8919	86.800	1.186
00057+4549	STT 547AE	0.8921	0.8952	86.800	2.364
00057+4549	POP 217AP	0.9048	0.9049	86.800	1.508
00251+1824	HJ 621	0.0773	0.0746	13.628	3.107
01057+3304	MLB 444	0.1019	0.1078	**3.617**	**3.473**
02516+4803	HJ 2160AB	0.0242	0.0250	1.073	2.456
03342+4837	BU 787AB	0.0382	0.0324	5.936	1.633
04312+5858	STI 2051AB	0.1043	0.1110	**181.244**	**181.273**
04556+1653	HJ 3263	0.0785	0.0754	2.828	0.839
05490+6342	TDS 197	0.2339	0.2215	1.371	11.177
05492+2941	BRT 2521	0.0888	0.0899	7.673	1.178
05535+4434	ES 1379	0.0607	0.0614	6.126	0.566
06092+6424	MLB 259	0.0378	0.0360	2.991	1.474
06277+2249	J 1092AB	0.0506	0.0547	1.537	4.056
07106+1543	J 703	0.0838	0.0847	12.162	2.751
08081+4916	HJ 2428	0.0736	0.0729	2.893	4.150
08089+1213	J 375AB	0.0523	0.0526	5.443	4.140
08503+0125	J 74	0.0545	0.0566	4.498	1.685
09388+0242	J 78	0.0742	0.0691	7.735	5.967
14031+1154	HJ 2699BC	0.1793	0.1897	12.212	6.890
14307+8308	LDS 1800	0.1022	0.1375	2.231	6.714
17046+3900	HJ 2804AB	0.0957	0.0945	5.578	3.490
18120+4355	ES 1419	0.0802	0.0794	4.338	1.371
18269+2950	HJ 1325	0.0152	0.0155	**0.546**	**0.423**
18489+1615	STF 2400AB	0.0735	0.0716	11.423	1.424
18580+6159	ES 1843	0.0356	0.0384	2.809	1.970
19289+3515	POP 34AB	0.1348	0.1536	16.822	0.380
19500+0637	J 1336AB	0.0442	0.0485	1.072	4.574
20087+1223	J 1338	0.0727	0.0753	0.318	5.839
22415+3256	J 2376AB	0.0667	0.0682	15.437	5.191
23581+2840	HJ 995	0.1135	0.1087	2.701	6.696

Then, we compared μ_{rel} with V for all 32 components in the linear solutions. As can be seen from Table 2, the values of V are in excellent agreement with μ_{rel} for all pairs. It is an argument in favor that the components are not gravitationally bound, i.e., that it is an optical pair. Another argument in favor of this is that the component parallaxes π_A and π_B are different. They are different for 29 pairs in our sample. For the remaining three pairs, which component parallaxes are approximately equal, additional criteria should be applied to determine the nature of their movements. These parallaxes are given in bold face in Table 2.

Table 3: The speed of relative motion μ, the characteristic speed of relative motion μ^* and the normalized relative motion μ′ for 32 pairs which are identified by their WDS numbers and discoverer's designations

WDS Design.	Discoverer	μ	μ^*	μ′
α,δ (2000)	Designation	("/yr)	("/yr)	
00057+4549	STT 547AC	0.93651	0.01607	58.3
00057+4549	STT 547AD	0.91173	0.01645	55.4
00057+4549	STT 547AE	0.93990	0.02252	41.7
00057+4549	POP 217AP	0.87604	0.04865	18.0
00251+1824	HJ 621	0.08390	0.00452	18.6
01057+3304	MLB 444	0.10376	0.00043	240.1
02516+4803	HJ 2160AB	0.03213	0.00007	450.8
03342+4837	BU 787AB	0.03704	0.00282	13.1
04312+5858	STI 2051AB	0.10175	0.13230	**0.8**
04556+1653	HJ 3263	0.08431	0.00026	327.8
05490+6342	TDS 197	0.29292	0.00017	1754.5
05492+2941	BRT 2521	0.10240	0.00197	52.0
05535+4434	ES 1379	0.06624	0.00141	46.9
06092+6424	MLB 259	0.03457	0.00048	72.0
06277+2249	J 1092AB	0.05374	0.00016	332.9
07106+1543	J 703	0.08686	0.00289	30.0
08081+4916	HJ 2428	0.07765	0.00054	142.8
08089+1213	J 375AB	0.05402	0.00089	60.6
08503+0125	J 74	0.05876	0.00069	85.8
09388+0242	J 78	0.06944	0.00316	22.0
14031+1154	HJ 2699BC	0.12684	0.00340	37.3
14307+8308	LDS 1800	0.12619	0.00068	185.8
17046+3900	HJ 2804AB	0.10571	0.00126	84.1
18120+4355	ES 1419	0.08356	0.00066	126.3
18269+2950	HJ 1325	0.02183	0.00004	567.0
18489+1615	STF 2400AB	0.07898	0.00361	21.9
18580+6159	ES 1843	0.03934	0.00047	83.2
19289+3515	POP 34AB	0.15468	0.00518	29.9
19500+0637	J 1336AB	0.04701	0.00010	456.3
20087+1223	J 1338	0.07717	0.00001	5479.4
22415+3256	J 2376AB	0.06813	0.00558	12.2
23581+2840	HJ 995	0.10480	0.00044	239.3

To answer the question concerning the nature of motion of these pairs we need observations covering very long time intervals. Otherwise, the only possibility is to apply existing criteria for establishing whether the pair is gravitationally bound or not. We applied the method given in Tokovinin and Kiyaeva (2016). The characteristic speed of relative motion μ^* was calculated by using formula 2 from this paper, and the speed of relative motion μ was calculated by using the quantities obtainable from formulae 4 and 5. The normalized relative motion $\mu' = \mu/\mu^*$ was used in that paper as the criterion of determining the motion nature. A physical pair has $\mu' < 1.5$. When the normalized relative motion is $\mu' > 1.5$, the double star is an optical pair.

Then, we can compare these speeds, μ and μ^*. Optical pairs usually move substantially faster than the characteristic speed. As can be seen in Table 3, relative motion μ is much higher than the characteristic speed of relative motion μ^* in all pairs except in the case of pair STI 2051 AB. Its the normalized relative motion $\mu' = 0.8$. As the parallaxes of both components are almost equal (see Table 2), this pair is most likely a long-period binary.

4. CONCLUSIONS

From 2004 until now we have obtained several thousand CCD frames of double and multiple stars at NAOR. The data were published and sent to the international databases. There is a small number of pairs for which the star images were not visually separated and the measurements could not be carried out. The reasons are the proximity of the components and the limiting capabilities of the CCD camera and seeing.

Using our observational data, we have calculated the first orbits for two binaries, recalculated three orbits and for the first time calculated linear solutions for 32 pairs. By applying the criteria for establishing the nature of the movement, it was confirmed that the components are not gravitationally bound in 31 pairs. Only one pair is most likely a long-period binary.

Acknowledgements

The results presented in many papers and in this poster are based on observations with the 2-m RCC telescope of Rozhen National Astronomical Observatory operated by Institute of Astronomy, Bulgarian Academy of Sciences. The authors gratefully acknowledge the observing grant support from the Institute of Astronomy and Rozhen National Astronomical Observatory, Bulgarian Academy of Sciences.

This research has made use of the Washington database maintained at the U.S. Naval Observatory and of the SIMBAD database, operated at CDS, Strasbourg, France.

This research has been supported by the Ministry of Education and Science of the Republic of Serbia, and these results are parts of the Grant 451-03-68/2022-14/200002 of Astronomical Observatory.

References

Cvetković Z., Novaković B., Strigachev A., Popović G.M.: 2006, *Serb. Astron. J.*, 172, 53-58.

Cvetković Z., Pavlović R., Strigachev A., Novaković B., Popović G.M.: 2007, *Serb. Astron. J.*, 174, 83-88.

Cvetković Z., Ninković S.: 2008, *New Astronomy*, 13, 587-592.

Cvetković Z., Pavlović R., Boeva S.: 2010, *Serb. Astron. J.*, 180, 103-107.

Cvetković Z., Pavlović R., Damljanović G., Boeva S.: 2011, *Astronomical Journal,* 142, 73, 9pp.

Cvetković Z., Pavlović R., Ninković S., Stojanović M.: 2012a, *Astronomical Journal,* 144, 80, 8pp.

Cvetković Z., Damljanović G., Pavlović R., Vince O., Milić I.S., Stojanović M.: 2012b, *Serb. Astron. J.*, 184, 97-104.

Cvetković, Z., Damljanović, G., Pavlović, R.: 2013, *Publications of the Astronomical Society "Rudjer Bošković,"* 12, 135-142.

Cvetković Z., Pavlović R., Boeva S.: 2015, *Astronomical Journal,* 149, 150, 9pp.

Cvetković Z., Pavlović R., Boeva S.: 2016, *Astronomical Journal,* 151, 58, 9pp.

Cvetković Z., Pavlović R., Boeva S.: 2017, *Astronomical Journal,* 153, 195, 11pp.

Cvetković Z., Pavlović R., Boeva S.: 2018, *Astronomical Journal,* 156, 134, 11pp.

Cvetković Z., Pavlović R., Boeva S.: 2019, *Astronomical Journal,* 158, 215, 10pp.

Cvetković Z., Pavlović R., Boeva S.: 2021, *Astronomical Journal,* 162, 223, 17pp.

Novaković B.: 2007, *Baltic Astronomy*, 16, 435-442.

Pavlović R., Cvetković Z., Olević D., Strigachev A., Popović G.M., Novaković B.: 2005, *Serb. Astron. J.*, 171, 49-53.

Pavlović R., Cvetković Z., Boeva S., Vince O., Stojanović M.: 2013, *Astronomical Journal,* 146, 52, 7pp.

Tokovinin, A., Kiyaeva, O.: 2016, *Monthly Notices of the Royal Astronomical Society*, 456, 2070-2079.

Proceedings of the XIII Bulgarian-Serbian Astronomical Conference
(XIII BSAC) Velingrad, Bulgaria, October 3-7, 2022
Editors: Evgeni Semkov, Milan S. Dimitrijević, Momchil Dechev and Zoran Simić
Publ. Astron. Soc. "Rudjer Bošković" No 25, 2023, 43-51

GAIA ALERTS AND BULGARIAN-SERBIAN COOPERATION FROM 2014 TO 2022

GORAN DAMLJANOVIĆ[1], MILAN STOJANOVIĆ[1],
RUMEN BACHEV[2] and SVETLANA BOEVA[2]

[1]*Astronomical Observatory, Volgina 7, 11060 Belgrade, Serbia*
[2]*Institute of Astronomy and NAO, Bulgarian Academy of Sciences, 72 Tsarigradsko Shosse, 1784 Sofia, Bulgaria*
E-mail: gdamljanovic@aob.rs, mstojanovic@aob.rs,
bachevr@astro.bas.bg, sboeva@astro.bas.bg

Abstract. We started to observe the Gaia Alerts or Gaia Fellow Up Network for Transients Objects (Gaia-FUN-TO) from October 2014. Because of better results, at 2013 we established the local cooperation "Serbian-Bulgarian mini-network telescopes" using 6 instruments. Our activities are in accordance with actual Serbian and Bulgarian Academies of Sciences (SANU-BAN) joint research project "Gaia Celestial Reference Frame (CRF) and fast variable astronomical objects" (during three years period, 2020-2022). There are two sites in Bulgaria (at Belogradchik and Rozhen) and one site in Serbia (Astronomical Station Vidojevica - ASV). Before the COVID-problem and during 2020, we did about 15 Gaia Alerts objects per year. During 2021, we observed just 5 Gaia Alerts (Gaia19dke, Gaia21awo, Gaia21azb, Gaia21efs, Gaia21ehu) but Gaia19dke 11 times and Gaia21azb 13 times; about 300 CCD images were done. Also, during 2022 (until 1st October 2022), we did 5 Gaia Alerts (Gaia21cgt, Gaia22aeu, Gaia22atp, Gaia22btj, Gaia22btc); about 50 CCD images. We present some results, here.

1. INTRODUCTION

Gaia astronomical satellite is surveying the full sky (astrometrically, photometrically, and spectroscopically), and it is a space mission of the European Space Agency (ESA) operating since mid-2014. Gaia results are useful for all the relevant scientific communities: our understanding of the Milky Way galaxy, for the stellar physics, the Solar system bodies, astrometry, etc. The high-precision astrometric data (the positions, proper motions and parallaxes) are the main goal of Gaia mission. The Gaia catalogue (for the objects from G = 3 mag to 21 mag) is an important step in the realization of the Gaia reference frame in future. The third

Gaia solution or data release (Gaia DR3) has been made publicly available on 13th June 2022 (there are about 1.806 billion sources with G magnitude).

The Gaia is scanning the sky multiple times and provides near-real-time photometric data. Because of that process and data, it is possible to detect some changes in brightness from all over the sky and appearance of new objects. As a result, the Gaia Science Alerts system produces alerts on these interesting objects. After that, using the ground-based telescopes (as the "Serbian-Bulgarian mini-network telescopes") we continue astronomical observations of the mentioned objects. The Gaia Photometric Science Alerts published the first alerts in October 2014. Just few years after that, the Gaia Science Alerts was among the leading transient surveys in the world (more than 3000 transients were discovered during three years). During 2022, about 4000 transients were discovered and it is really impressive result. The transients are: supernovae, cataclysmic variables, microlensing events, other rare phenomena.

2. SERBIAN-BULGARIAN COOPERATION IN LINE WITH GAIA ALERTS

In line with the plan of astronomical observations, Gaia is recording each object in the sky about 70 times over the mission lifetime. Because of it, there are a lot of alerts. The number of alerts increases with number of observations of each object, and during 2022 it had been reported about 4000 Gaia Alerts from all over the sky. Some numbers of Gaia Alerts (published per year) are possible to find on the official website[1].

The Astronomical Station Vidojevica (ASV) of Astronomical Observatory in Belgrade (AOB) was established in 2011, and it was the Serbian new astronomical site. At the beginning we used a new D=60 cm telescope "Nedeljkovic", and at mid-2016 there was another new telescope (D=1.4 m) "Milankovic" via the Belissima project (see website[2]). In Bulgaria (at Belogradchik and Rozhen sites), from 2013 we used 4 instruments in line with our regional collaboration "Serbian-Bulgarian mini-network telescopes". Mentioned collaboration is in line with the Serbian Academy of Sciences and Arts (SANU) and Bulgarian Academy of Sciences (BAN) or SANU-BAN joint research projects: "Observations of ICRF radio-sources visible in optical domain" (for three years period, 2014-2016), "Study of ICRF radio-sources and fast variable astronomical objects" (2017-2019), "Gaia Celestial Reference Frame (CRF) and fast variable astronomical objects" (2020-2022), and the next one "Gaia astrometry and fast variable astronomical objects" (2023-2025); the leaders are G.Damljanović (Serbia) and R.Bachev (Bulgaria). Some information about the telescopes and CCD cameras were published in a few papers (Damljanović et al. 2014; Taris et al. 2018; Damljanović et al. 2020); see Table 1. The first column in Table 1 is: telescope

1 http://gsaweb.ast.cam.ac.uk/alerts/home

2 http://belissima.aob.rs

(with diameter - D and focal length – F in meters), and site (ASV, Rozhen or Belogradchik). The other columns are: CCD camera, chip size (in pixels), pixel size (in micrometers), scale (in arcseconds), and field of view – FoV (in arcminutes).

Table 1: Telescopes and CCD cameras of Serbian-Bulgarian cooperation.

Telescope D/F [m]	*Camera*	*Chip size [pixel]*	*Pixel size [μm]*	*Scale ["]*	*Field of view - FoV[']*
1.4/11.42 ASV	Apogee Alta U42 Andor iKon-L	2048 x 2048 2048 x 2048	13.5 x 13.5 13.5 x 13.5	0.243 0.24	8.3 x 8.3 8.3 x 8.3
0.6/6 ASV	Apogee Alta U42 FLI PL230	2048 x 2048 2048 x 2064	13.5 x 13.5 15 x 15	0.465 0.517	15.8 x 15.8 17.7 x 17.8
2/15.774 Rozhen	VersArray 1300B Andor iKon-L	1340 x 1300 2048 x 2048	20 x 20 13.5 x 13.5	0.261 0.176	5.6 x 5.6 6.0 x 6.0
0.6/7.5 Rozhen	FLI PL09000	3056 x 3056	12 x 12	0.33	16.8 x 16.8
0.6/7.5 Belogradchik	FLI PL09000	3056 x 3056	12 x 12	0.33	16.8 x 16.8
0.5/0.7/1.72 Rozhen	FLI PL16803	4096 x 4096	9 x 9	1.08	73.7 x 73.7

About Bulgarian telescopes, three instruments are at National Astronomical Observatory (NAO) BAN Rozhen (D=2 m, D=60 cm, and Schmidt-camera 50/70 cm), and one instrument is at Belogradchik AO site (D=60 cm). Soon, we hope to use the MEADE D=40 cm instrument at ASV (see Figure 1) and new D=1.5 m telescope at Rozhen NAO BAN (see Figure 2, Dr. R. Pavlovic is on that figure).

3. RESULTS

There are a few published papers (Damljanovic et al. 2014; Campbell et al. 2015; Wyrzykowski et al. 2020; Szegedi-Elek et al. 2020; Damljanovic et al. 2020; Hodgkin et al. 2021; etc.) using our data obtained from six telescopes at three sites; it is in line with “Serbian-Bulgarian mini network telescopes” and mentioned SANU-BAN joint research projects. Also, we can use D=1.31 m telescope of the Aryabhatta Research Institute of observational sciencES (ARIES) in accordance with our useful cooperation with few colleagues from India (A.C. Gupta and others). Usually, we did 3 CCD images per filter, and the Johnson-Cousins BVRcIc filters were available. After that, the standard bias, dark and flat-fielded corrections are done (also, hot/dead pixels are removed), and the Astrometry.Net and Source Extractor are used. For further calibration, the output is supposed to be submitted to the Cambridge Photometric Calibration Server (CPCS). During Oct. 2014 – Oct. 2022, we collected about 3700 CCDs for ~100 Gaia Alerts or Gaia-Follow-Up Network for Transients Objects (Gaia-FUN-TO), and it was ~550 images per year (near 15 objects per year) before the problem with COVID virus.

Table 2: Photometry results of Gaia19dke (from 29th Aug.2019 to 31st Oct.2021).

Date	*JD [day]* *Filter [mag]*	*JD [day]* *Filter [mag]*	*JD [day]* *Filter [mag]*	*JD [day]* *Filter [mag]*
29Aug.2019	2458725.42 B=17.49±0.05	2458725.43 V=15.75±0.03	2458725.42 J=13.67±0.02	2458725.44 i=14.65±0.01
12Nov.2020	2459166.20 B=17.35±0.02	2459166.23 V=15.49±0.01		2459166.24 i=14.59±0.01
19Nov.2020	2459173.22 B=17.46±0.02	2459173.21 V=15.74±0.01		2459173.22 i=14.58±0.01
17Dec.2020	2459201.20 B=17.44±0.03	2459201.19 V=15.69±0.01		2459201.19 i=14.63±0.01
20Dec.2020	2459204.18 B=17.36±0.04	2459204.18 V=15.50±0.01		2459204.19 i=14.57±0.01
18Feb.2021	2459264.66 B=17.49±0.03	2459264.63 V=15.84±0.01		2459264.67 i=14.62±0.01
10May2021	2459345.47 B=17.17±0.01	2459345.47 V=15.45±0.01	2459345.46 J=13.24±0.01	2459345.50 r=15.22±0.01
11May2021	2459346.50 B=17.55±0.03	2459346.48 V=15.62±0.01		2459346.48 i=14.62±0.01
12May2021	2459347.45 B=17.28±0.13	2459347.48 V=15.59±0.03		2459347.46 i=14.61±0.02
10June2021	2459376.48 B=17.40±0.08	2459376.47 V=15.69±0.03		2459376.48 i=14.63±0.01
10June2021	2459377.47 B=17.44±0.02	2459377.48 V=15.58±0.01		2459377.48 i=14.62±0.01
11July2021	2459407.38 B=17.39±0.01	2459407.39 V=15.69±0.01	2459407.38 J=13.31±0.01	2459407.39 i=14.45±0.01
12July2021	2459408.38 B=17.44±0.01	2459408.38 V=15.68±0.01		2459408.37 i=14.46±0.01
8Aug.2021	2459435.31 B=17.48±0.01	2459435.31 V=15.73±0.01		2459435.31 i=14.59±0.01
12Sep.2021	2459470.27 B=17.57±0.01	2459470.27 V=15.74±0.01		2459470.27 i=14.58±0.01
31Oct.2021	2459519.21 B=17.36±0.01	2459519.21 V=15.71±0.01	2459519.21 J=13.44±0.01	2459519.21 i=14.65±0.01

During 2021, we did 5 objects: Gaia19dke (10 times), Gaia21awo, Gaia21azb (13 times), Gaia21ehu, Gaia21efs; near 350 CCDs. Also, during 2022 (until 1st October), we did 5 objects: Gaia21cgt, Gaia22aeu, Gaia22atp, Gaia22btj, Gaia22btc; near 60 CCDs.

Figure 1: The MEADE 40 cm instrument at ASV (Serbia).

Last few years, the objects **Gaia18dvy** ($\alpha=20^h{:}05^m{:}06.^s02$, $\delta=36^o{:}29'{:}13.''52$) and **Gaia19dke** ($\alpha$=19:25:58.68, δ=28:24:24.70, see Figure 3) were very interesting. Using two ASV instruments (60 cm and 1.4 m), the Gaia18dvy was observed three times during 2019: on 12th March (JD=2458555.6) using D=1.4 m ASV telescope with CCD Andor iKon-L (0.″244 per pixel, there are 12 CCD images or 3(BVRI)), on 29th March (JD=2458572.6) using the same instrument to get 3(BVRI), and on 29th August (JD=2458725.5) using D=60 cm with FLI PL230 (0.″518, 3(BVRI)).

Figure 2: The concrete pillar of new 1.5 m telescope (in November 2022) at Rozhen NAO BAN (Bulgaria).

Mentioned data were useful for a paper about the Gaia18dvy as a new case of FU Orionis-type young eruptive stars in the Cygnus OB3 Association (Szegedi-Elek et al. 2020). Some co-authors of that paper are few our colleagues (in line with the Serbian-Bulgarian cooperation): S.Boeva and G.Latev from Bulgaria, O.Vince, G.Damljanovic and M.D.Jovanovic from Serbia, and A.C.Gupta from India.

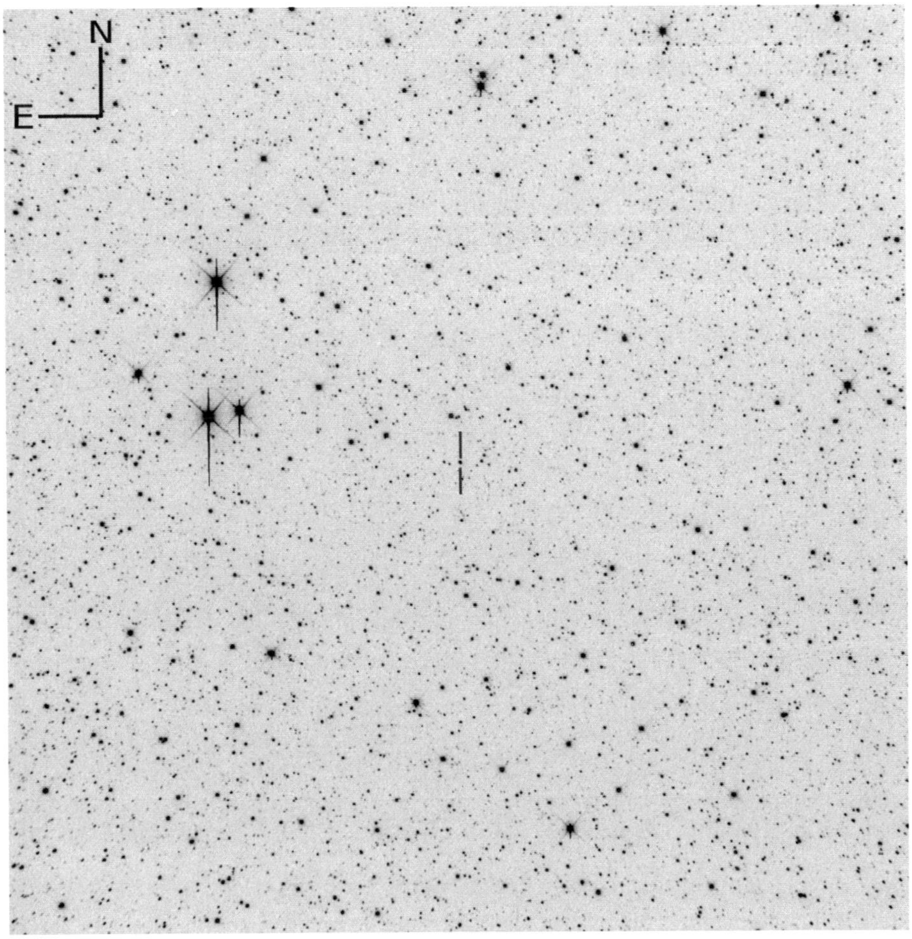

Figure 3: Our image (marked with vertical lines) of the Gaia19dke (R-filter, Exp.=120s) on October 31st 2021 using the D=1.4 m ASV telescope with CCD Andor iKon-L camera.

About Gaia19dke, our original CCD image of that object is presented (see Figure 3) after standard reduction (bias/dark/flat, hot/dead pixels, etc.), and the object is marked with vertical lines. The east is left and north is up. This image was made on 31st October 2021 using the ASV telescope D=1.4 m with CCD Andor iKon-L camera: R-filter, Exp.=120s, FoV=8.′3x8.′3, binning=1x1, scale=0.″391 per pixel. Our obtained results of Gaia19dke are in Figure 4 and Table 2: the first column is date of observation, other 4 columns are Julian Date - JD and magnitudes (with standard deviation) after photometry (using suitable filters). The standard deviation of our magnitudes is on the level 0.01 mag. There are 16 dates from 29th August 2019 to 31st October 2021 (about two years of observations). Light curve of data is in Figure 4: x-axis is t=JD-2450000 (in days), and y-axis is suitable magnitude. Presented results are in good accordance with: the ground-based relative photometry, results obtained from other telescopes (at site https://cpcs.astrolabs.pl/cgi/upload), possibilities of our instruments, etc. These magnitudes are transferred from our set of filters (Johnson

BV and Cousins RcIc) into another one via the Cambridge Server (into J, r, and i using the APASS, 2MASS and PS1 catalogs).

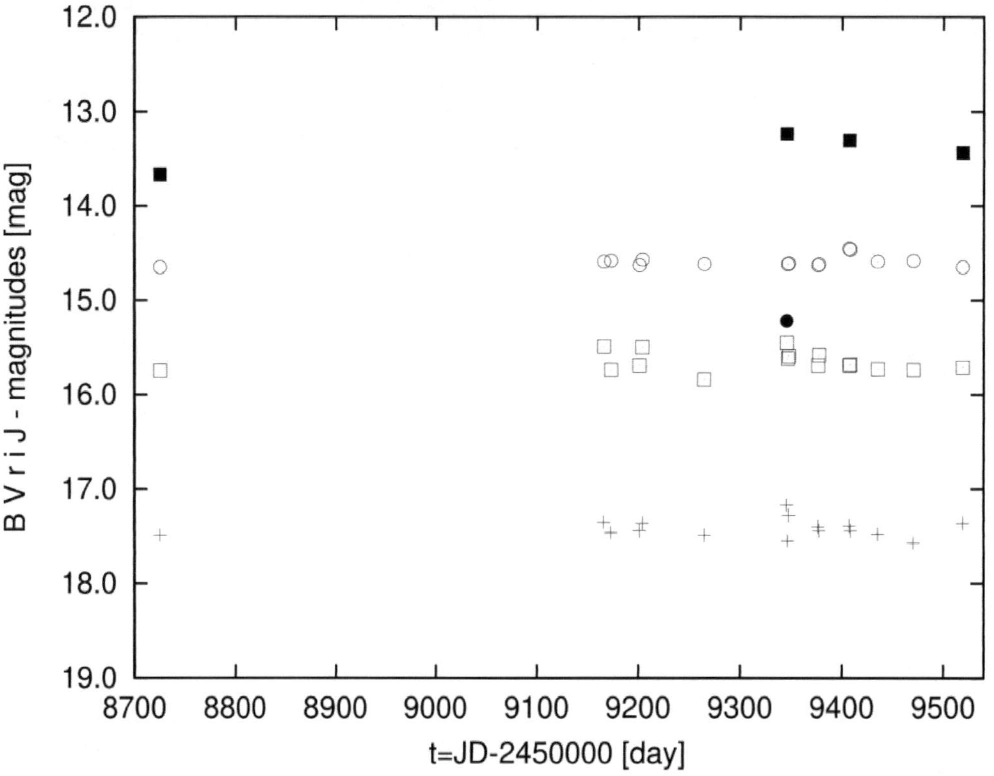

Figure 4: Light curve of Gaia19dke using our data from August 29th 2019 to October 31st 2021: B (cross), V (white rectangle), r (black circle), i (white circle), and J (black rectangle) magnitudes with time t.

Once identified, Gaia19dke was observed by Follow-Up-Network of telescopes in multi-national campaign. All follow-up data were collected and presented in the light curve of Gaia19dke (at site https://cpcs.astrolabs.pl/cgi/upload). Our data (16 points in B and V filters, etc.) are among the others collected for this campaign and they fit very well other points of the light curve. We hope, there are enough relevant Gaia19dke data for our next paper in the future.

4. CONCLUSIONS

The first observations of the ESA Gaia astronomical satellite were done in mid-2014. Since October 2014, the Gaia Photometric Science Alerts started to publish alerts, and there were ~3000 alerts during the first three years. They were issued by the Gaia Science Alerts group: cataclysmic variables, supernovae, candidate microlensing events, etc. About 4000 transients were discovered (as really impressive result) during 2022. To be a part of that interesting project, we established our regional cooperation (the "Serbian-Bulgarian mini-network telescopes") and few SANU-BAN projects. We used 6 Serbian-Bulgarian telescopes for the mentioned activities from mid-2016. Using D=2 m Rozhen and D=1.4 m ASV telescopes (with Exp.=300s) it is possible to observe objects down to 20 mag in the V-band (down to 19 mag using smaller telescopes). The MEADE 40 cm telescope at ASV is a new addition to our network from the end of 2022, and we hope to use the 1.5 m instrument at Rozhen NAO BAN, soon. Until the October 1st 2022, we observed about 100 Gaia Alerts (usually, near 15 objects per year) or near 3700 CCD images (~550 images per year). These observations (usually, three CCD images per filter) were done in Johnson BV and Cousins RcIc filters. Few papers were published with our results. Three papers were published during 2020 and 2021: (Wyrzykowski et al. 2020) about Gaia16aye or Ayers Rock, (Szegedi-Elek et al. 2020) about Gaia18dvy, and (Hodgkin et al. 2021) about the main results of Gaia Alerts. Some of our results were presented at a few conferences. As an example, the object Gaia18dvy is a new case of FU Orionis-type young eruptive stars in the Cygnus OB3 Association. We hope, the Gaia19dke (which we did two years) will be the subject of next paper, soon.

Acknowledgements

We gratefully acknowledge the observing grant support from the Institute of Astronomy and Rozhen NAO (BAS), and support by the Ministry of Education, Science and Technological Development of the Republic of Serbia (contract No 451-03-68/2022-14/200002). We acknowledge ESA Gaia DPAC and Photometric Science Alerts Team (http://gsaweb.ast.cam.ac.uk/alerts).

References

Damljanović G., Vince O., Boeva S.: 2014, *Serb. Astron. J.*, 188, 85-93.

Damljanović G., Boeva S., Latev G., Bachev R., Vince O., Jovanović M.D., Cvetković Z., Pavlović R.: 2020, *Bulgarian Astron. J.*, 32, 108-112.

Campbell H.C., Marsh T.R., Fraser M., …, Damljanović G., …, Vince O., et al.: 2015, *MNRAS*, 452, 1060-1067.

Szegedi-Elek E., Abraham P., Wyrzykowski L, …, Boeva S, …, Damljanović G., …, Gupta A.C., …, Jovanović M.D., …, Latev, G., …, Vince O., et al.: 2020, *ApJ*, 899, 130(1-8).

Taris F., Damljanović G., Andrei A., et al.: 2018, *A&A*, 611, A52.

Wyrzykowski L., Mroz P., Rybicki K.A., …, Damljanović G., et al.: 2020, *A&A*, 633, A98.

Hodgkin S.T., Harrison D.L., Breedt E., …, Damljanović G., et al.: 2021, *A&A*, 652, A76.

Proceedings of the XIII Bulgarian-Serbian Astronomical Conference
(XIII BSAC) Velingrad, Bulgaria, October 3-7, 2022
Editors: Evgeni Semkov, Milan S. Dimitrijević, Momchil Dechev and Zoran Simić
Publ. Astron. Soc. "Rudjer Bošković" No 25, 2023, 53-62

STARK BROADENING OF Fe XXV LINES FOR NEUTRON STARS AND THEIR ENVIRONMENT INVESTIGATIONS

MILAN S. DIMITRIJEVIĆ[1,2], MAGDALENA D. CHRISTOVA[3], CRISTINA YUBERO[4] and SYLVIE SAHAL-BRÉCHOT[2]

[1]*Astronomical Observatory, Volgina 7, 11060 Belgrade, Serbia*
[2]*LERMA, Observatoire de Paris, Université PSL, CNRS, Sorbonne Université, 5 Place Jules Janssen, 92190 Meudon, France*
[3]*Department of Applied Physics, FAMI, Technical University of Sofia, blvd. Kl. Ohridski 10, 1000 Sofia, Bulgaria*
[4]*Departamento de Física, Campus de Rabanales Edif. C2, Universidad de Córdoba, E-14071 Córdoba, Spain*
E-mail: mdimitrijevic@aob.bg.ac.rs, mchristo@tu-sofia.bg, f62yusec@uco.es, sylvie.sahal-brechot@obspm.fr

Abstract. Stark broadening parameters, full widths at half intensity maximum and shifts for important spectral lines of Fe XXV, broadened by electron-, proton- and Fe XXVII ions-impacts, have been calculated with the help of the semiclassical perturbation formalism, for plasma conditions of interest for neutron star atmospheres and their environment, as well as for inertial fusion plasma. Examples of obtained results and the corresponding discussion are presented.

1. INTRODUCTION

Stark broadening of spectral lines, or broadening by collisions with charged particles is significant for many important topics in astrophysics as for example modelling of stellar plasma, analysis and synthesis of spectral lines, opacity, radiative transfer, abundance determination, gravity acceleration etc. In the conditions of neutron star atmospheres and their envirionment Stark broadening is the most important pressure broadening mechanism of spectral lines.

In spite of importance of Stark broadening parameters for analysis and modelling of neutron star atmospheres and their environements, we can see in literature that they are calculated very approximately, and without taking into account the magnetic field (see e.g. Paerels, 1997). Madej (1989) and Majczyna et al. (2005), use for Stark broadening calculations in atmospheres of neutron stars approximate formula from Chap. IV 6 of Griem's (1974) book, while Suleimanov et al. (2014) use very approximate formula of Cowley (1971). Moreover, Werner et al. (2007) performed synthesis of spectrum of neutron stars neglecting Stark broadening of considered lines. In their synthetic spectrum are present numerous lines of iron ions, from Fe XVII up to Fe XXVI, including Fe XXV. Also, Cottam et al. (2002) found in X-ray burst spectra of EXO 0748–676, a Fe XXV feature. One could be seen in Van Peet's et al. (2009) analysis of environments of neutron stars, that there exist places where electron density and temperature are favorable for Stark broadening. It should be noted as well, that Stark broadening of higly charged iron ions may be of interest and for diagnostics of plasma during attempts to create neutron star plasma conditions in laboratory (see e.g. Moon et al., 2005).

In order to provide reliable Stark broadening parameters needed for investigations of neutron stars, we calculated widths and shifts of 18 Fe XXV spectral lines, broadened by collisions with important charged constituents of neutron star atmospheres, electrons, protons and Fe XXVII ions. Calculations have been performed for plasma conditions of interest for neutron star atmospheres and their environments. The complete results are published in Dimitrijević et al. (2023), and here we present an example of the obtained results and discuss it.

2. THEORY

In order to perform calculations of helium-like Fe XXV Stark full widths at half intensity maximum (FWHM) and shifts, the semiclassical perturbation theory (Sahal-Bréchot, 1969ab) has been used. We note that the later inovations and opimizations may be found in Sahal-Bréchot (1974, 1991), Fleurier et al. (1977), Dimitrijević et al. (1991), Dimitrijević and Sahal-Bréchot (1996) and Sahal-Bréchot et al. (2014). Acording to the semiclassical perturbation method, FWHM - W and shift - d of an isolated spectral line of a non-hydrogenic ion are:

$$W=2n_e\int_0^{\infty} \mathrm{v}f(v)\mathrm{dv}\left[\sum_{i'\neq i}\sigma_{ii'}(v)+\sum_{f'\neq f}\sigma_{ff'}(v)+\sigma_{el}\right]$$

$$d=\int_0^{\infty} \mathrm{v}f(v)\mathrm{dv}\int_{R_3}^{R_d} 2\pi\rho\, d\rho \sin 2\phi_p$$

Here, i and f denote the initial and final level of the corresponding transition; i ' and f ' are perturbing levels; n_e is electron density; v perturber velocity, f(v) the Maxwellian distribution of electron velocities, and $\sigma_{ii'}(v)$, $\sigma_{ff'}(v)$ the cross sections for inelastic collisions. These cross sections may be calculated using an integration of the transition probability $P_{ii'}$ (respectively $P_{ff'}$), over the impact parameter ρ:

$$\sum_{i'\neq i}\sigma_{ii'}(v)=\frac{1}{2}\pi R_1^2+\int_{R_1}^{R_d}2\pi\rho\,d\rho\sum_{i'\neq i}P_{ii'}(\rho,v)$$

The cross section for elastic collisions is expressed as:

$$\sigma_{el}=2\pi R_2^2+\int_{R_2}^{R_d}8\pi\rho\,d\rho\sin^2\delta+\sigma_r$$

$$\delta=\left(\phi_p^2+\phi_q^2\right)^{1/2}$$

In the above equations, δ is the phase shift due to polarization (φ_p) and quadrupole (φ_q,) potentials. The procedure for cut-off parameters R_1, R_2, R_3 and the Debye cut-off R_d are explained in Section 1, Chapter 3 in Sahal-Bréchot (1969b). The term σ_r accounts for Feshbach resonances, described in Fleurier et al. (1977) and Sahal-Bréchot (2021). Since the Coulomb force is not attractive but repulsive for perturbing ions, in this case trajectories are different, but formulae are analogous.

3. RESULTS AND DISCUSSION

With the help of the semiclassical perturbation theory (Sahal-Bréchot, 1969ab), we calculated electron-, proton-, and Fe XXVII ion-impact broadening parameters, FWHM - W and shift - d for 18 spectral lines of helium-like Fe XXV, for temperatures within the interval 300 000 K - 20 000 000 K and for perturber densities from 10^{17} to 10^{24} cm^{-3}. Energy levels are from Sugar and Corliss (1985), Shirai et al. (2000) and Kramida et al. (2021). In order to obtain oscillator strengths we used the method of Bates and Damgaard (1949) together with tables of Oertel and Shomo (1968). When these tables are not applicable, for transitions involving higher energy levels, the reference Van Regemorter et al. (1979) has been used for calculations of needed oscillator strengths.

The results for Stark FWHM and shift for 18 Fe XXV spectral lines broadened by electron-, proton- and Fe XXVII ion-impacts, for all investigated temperatures and electron densities are published in Dimitrijević et al. (2023). Here, as an example of obtained results, we present in Table 1 the results for 18 Fe XXV spectral lines, broadened by collisions with electrons and Fe XXVII ions, important perturbers in atmospheres of neutron stars, for an electron density of 10^{19} cm^{-3}.

Table 1: This table gives electron-, and Fe XXVII ion-impact broadening parameters for Fe XXV lines. The wavelengths of the spectral lines (in Å), are theoretically determined, and the quantity C (Dimitrijević and Sahal-Bréchot, 1984) are provided as well. If this quantity is divided with Stark width, one obtains maximal pertuber density for which one can take that the line is isolated. The Stark broadening parameters for 18 Fe XXV spectral lines are given for perturber density of 10^{19} cm^{-3}. The sign of the shift is positive if the line is shifted in direction of the higher wavelengths. With an asterisk are denoted values which are near the limit of validity of the impact approximation.

PERTURBER DENSITY = 10^{19} cm^{-3}

TRANSITION	T(K)	ELECTRONS WIDTH(A)	ELECTRONS SHIFT(A)	Fe XXVII WIDTH(A)	Fe XXVII SHIFT(A)
SINGLETS					
Fe XXV $1s^{2}\,{}^{1}S-2p\,{}^{1}P^{o}$	300000.	0.656E-07	-0.244E-07	0.242E-11	-0.451E-10
1.9 Å	500000.	0.460E-07	-0.984E-08	0.126E-10	-0.297E-09
C = 0.89E+17	1000000.	0.359E-07	-0.325E-08	0.489E-10	-0.123E-08
	5000000.	0.175E-07	-0.639E-10	0.689E-09	-0.881E-08
	10000000.	0.125E-07	-0.474E-10	0.276E-08	-0.179E-07
	20000000.	0.897E-08	-0.316E-10	0.113E-07	-0.338E-07
Fe XXV $1s^{2}\,{}^{1}S-3p\,{}^{1}P^{o}$	300000.	0.182E-06	-0.187E-07	*0.149E-09	*0.561E-08
1.6 Å	500000.	0.141E-06	-0.704E-08	*0.107E-08	*0.369E-07
C = 0.26E+16	1000000.	0.101E-06	-0.169E-08	*0.975E-08	*0.153E-06
	5000000.	0.495E-07	0.490E-09	0.424E-06	0.830E-06
	10000000.	0.365E-07	0.404E-09	0.828E-06	0.119E-05
	20000000.	0.272E-07	0.283E-09	0.136E-05	0.149E-05
Fe XXV $2s\,{}^{1}S-2p\,{}^{1}P^{o}$	300000.	0.298E-02	-0.156E-03	0.103E-06	-0.807E-05
382.8 Å	500000.	0.214E-02	-0.739E-04	0.563E-06	-0.531E-04
C = 0.38E+22	1000000.	0.167E-02	-0.355E-04	0.238E-05	-0.220E-03
	5000000.	0.831E-03	-0.152E-04	0.106E-03	-0.157E-02
	10000000.	0.602E-03	-0.143E-04	0.652E-03	-0.306E-02
	20000000.	0.439E-03	-0.124E-04	0.232E-02	-0.517E-02

Table1: Continuation

Fe XXV $2s^1S$-$3p^1P^o$	300000.	0.779E-05	-0.156E-06	*0.613E-08	*0.232E-06
10.2 Å	500000.	0.605E-05	-0.497E-07	*0.440E-07	*0.153E-05
C = 0.11E+18	1000000.	0.438E-05	0.264E-08	*0.399E-06	*0.632E-05
	5000000.	0.215E-05	0.118E-07	0.175E-04	0.345E-04
	10000000.	0.159E-05	0.831E-08	0.344E-04	0.493E-04
	20000000.	0.119E-05	0.403E-08	0.564E-04	0.621E-04
Fe XXV $3s^1S$-$3p^1P^o$	300000.	0.141	-0.168E-02	*0.650E-04	*0.240E-02
1302.1 Å	500000.	0.110	-0.961E-03	*0.422E-03	*0.158E-01
C = 0.18E+22	1000000.	0.804E-01	-0.889E-03	0.318E-02	0.655E-01
	5000000.	0.406E-01	-0.892E-03	0.160	0.382
	10000000.	0.305E-01	-0.792E-03	0.365	0.564
	20000000.	0.231E-01	-0.670E-03	0.654	0.730
Fe XXV $2p^1P^o$-$3s^1S$	300000.	0.325E-05	0.630E-07	0.652E-09	0.967E-07
10.6 Å	500000.	0.243E-05	0.666E-07	0.530E-08	0.636E-06
C = 0.86E+17	1000000.	0.190E-05	0.888E-07	0.591E-07	0.264E-05
	5000000.	0.102E-05	0.832E-07	0.569E-05	0.163E-04
	10000000.	0.771E-06	0.722E-07	0.153E-04	0.249E-04
	20000000.	0.587E-06	0.581E-07	0.277E-04	0.342E-04

TRIPLETS

Fe XXV $2s^3S_1$-$2p^3P^o_0$	300000.	0.358E-02	0.218E-03	0.119E-06	-0.910E-05
428.2 Å	500000.	0.262E-02	0.716E-04	0.649E-06	-0.599E-04
C = 0.43E+22	1000000.	0.204E-02	0.461E-05	0.272E-05	-0.248E-03
	5000000.	0.101E-02	-0.178E-04	0.112E-03	-0.177E-02
	10000000.	0.731E-03	-0.169E-04	0.701E-03	-0.347E-02
	20000000.	0.533E-03	-0.149E-04	0.254E-02	-0.591E-02
Fe XXV $2s^3S_1$-$2p^3P^o_1$	300000.	0.313E-02	0.191E-03	0.105E-06	-0.780E-05
400.3 Å	500000.	0.229E-02	0.628E-04	0.568E-06	-0.513E-04
C = 0.37E+22	1000000.	0.179E-02	0.414E-05	0.238E-05	-0.213E-03
	5000000.	0.883E-03	-0.154E-04	0.952E-04	-0.152E-02
	10000000.	0.639E-03	-0.147E-04	0.597E-03	-0.298E-02
	20000000.	0.466E-03	-0.130E-04	0.216E-02	-0.508E-02

Table1: Continuation

Fe XXV $2s^3S_1$-$2p^3P^o_2$	300000.	0.144E-02	0.889E-04	0.493E-07	-0.325E-05
271.1 Å	500000.	0.106E-02	0.295E-04	0.267E-06	-0.214E-04
C = 0.17E+22	1000000.	0.825E-03	0.196E-05	0.110E-05	-0.887E-04
	5000000.	0.408E-03	-0.686E-05	0.387E-04	-0.635E-03
	10000000.	0.295E-03	-0.656E-05	0.241E-03	-0.125E-02
	20000000.	0.215E-03	-0.585E-05	0.887E-03	-0.214E-02
Fe XXV $2s^3S_1$-$3p^3P^o_0$	300000.	0.731E-05	-0.133E-06	0.236E-08	0.228E-09
10.0 Å	500000.	0.565E-05	-0.571E-07	0.126E-07	0.150E-08
C = 0.65E+18	1000000.	0.405E-05	-0.167E-07	0.489E-07	0.621E-08
	5000000.	0.196E-05	0.140E-08	0.564E-06	0.445E-07
	10000000.	0.145E-05	0.180E-08	0.141E-05	0.911E-07
	20000000.	0.108E-05	0.151E-08	0.278E-05	0.182E-06
Fe XXV $2s^3S_1$-$3p^3P^o_1$	300000.	0.730E-05	-0.131E-06	0.237E-08	0.501E-08
10.0 Å	500000.	0.564E-05	-0.556E-07	0.127E-07	0.330E-07
C = 0.70E+18	1000000.	0.405E-05	-0.143E-07	0.491E-07	0.137E-06
	5000000.	0.196E-05	0.230E-08	0.604E-06	0.977E-06
	10000000.	0.145E-05	0.222E-08	0.162E-05	0.191E-05
	20000000.	0.108E-05	0.176E-08	0.346E-05	0.326E-05
Fe XXV $2s^3S_1$-$3p^3P^o_2$	300000.	0.729E-05	-0.122E-06	0.249E-08	0.370E-07
10.0 Å	500000.	0.564E-05	-0.448E-07	0.136E-07	0.243E-06
C = 0.34E+16	1000000.	0.405E-05	-0.517E-08	0.590E-07	0.101E-05
	5000000.	0.197E-05	0.798E-08	0.196E-05	0.678E-05
	10000000.	0.145E-05	0.469E-08	0.616E-05	0.114E-04
	20000000.	0.108E-05	0.360E-08	0.118E-04	0.161E-04
Fe XXV $3s^3S_1$-$3p^3P^o_0$	300000.	0.198	-0.324E-02	0.652E-04	-0.175E-02
1552.8 Å	500000.	0.155	-0.167E-02	0.377E-03	-0.115E-01
C = 0.16E+23	1000000.	0.112	-0.158E-02	0.205E-02	-0.477E-01
	5000000.	0.548E-01	-0.145E-02	0.107	-0.301
	10000000.	0.410E-01	-0.126E-02	0.291	-0.470
	20000000.	0.311E-01	-0.101E-02	0.521	-0.650

Table1: Continuation

Fe XXV $3s^3S_1$-$3p^3P^o_1$	300000.	0.172	-0.278E-02	0.321E-03	-0.935E-02
C = 0.14E+23	1000000.	0.973E-01	-0.133E-02	0.169E-02	-0.388E-01
	5000000.	0.478E-01	-0.125E-02	0.861E-01	-0.247
	10000000.	0.358E-01	-0.109E-02	0.239	-0.389
	20000000.	0.271E-01	-0.875E-03	0.425	-0.538
Fe XXV $3s^3S_1$-$3p^3P^o_2$	300000.	0.754E-01	-0.113E-02	0.227E-04	-0.327E-03
956.0 Å	500000.	0.590E-01	-0.533E-03	0.124E-03	-0.215E-02
C = 0.31E+22	1000000.	0.426E-01	-0.504E-03	0.534E-03	-0.891E-02
	5000000.	0.210E-01	-0.490E-03	0.172E-01	-0.601E-01
	10000000.	0.157E-01	-0.452E-03	0.542E-01	-0.102
	20000000.	0.119E-01	-0.364E-03	0.105	-0.143
Fe XXV $2p^3P^o_0$-$3s^3S_1$	300000.	0.336E-05	-0.168E-06	0.513E-09	0.831E-07
10.4 Å	500000.	0.240E-05	0.930E-08	0.415E-08	0.547E-06
C = 0.69E+18	1000000.	0.177E-05	0.758E-07	0.461E-07	0.227E-05
	5000000.	0.926E-06	0.754E-07	0.477E-05	0.142E-04
	10000000.	0.711E-06	0.669E-07	0.133E-04	0.220E-04
	20000000.	0.542E-06	0.540E-07	0.240E-04	0.303E-04
Fe XXV $2p^3P^o_1$-$3s^3S_1$	300000.	0.558E-05	0.164E-06	*0.488E-08	*0.274E-06
10.4 Å	500000.	0.421E-05	0.172E-06	*0.413E-07	*0.180E-05
C = 0.69E+18	1000000.	0.317E-05	0.241E-06	*0.466E-06	*0.744E-05
	5000000.	0.178E-05	0.230E-06	0.208E-04	0.396E-04
	10000000.	0.141E-05	0.201E-06	0.399E-04	0.566E-04
	20000000.	0.110E-05	0.162E-06	0.639E-04	0.704E-04
Fe XXV $2p^3P^o_2$-$3s^3S_1$	300000.	0.574E-05	0.166E-06	*0.499E-08	*0.280E-06
10.5 Å	500000.	0.433E-05	0.176E-06	*0.422E-07	*0.184E-05
C = 0.71E+18	1000000.	0.326E-05	0.247E-06	0.213E-04	0.405E-04
	10000000.	0.145E-05	0.206E-06	0.408E-04	0.579E-04
	20000000.	0.113E-05	0.166E-06	0.654E-04	0.721E-04

The validity of the impact approximation is checked calculating the product of the collision volume V and the perturber density N. When NV< 0.1, the impact approximation is valid (Sahal-Bréchot, 1969ab). When 0.1<NV≤0.5, befor the corresponding values an asterisk is placed, to draw attention that such values are on the limit of validity of impact approximation. If the impact approximation is not valid (NV>0.5), Stark broadening parameters may be calculated with the help of the quasi-static approach (Griem 1974 or Sahal-Bréchot 1991) or, if and quasi-static approximation is not valid, by using a unified method as for example Barnard et al. (1974).

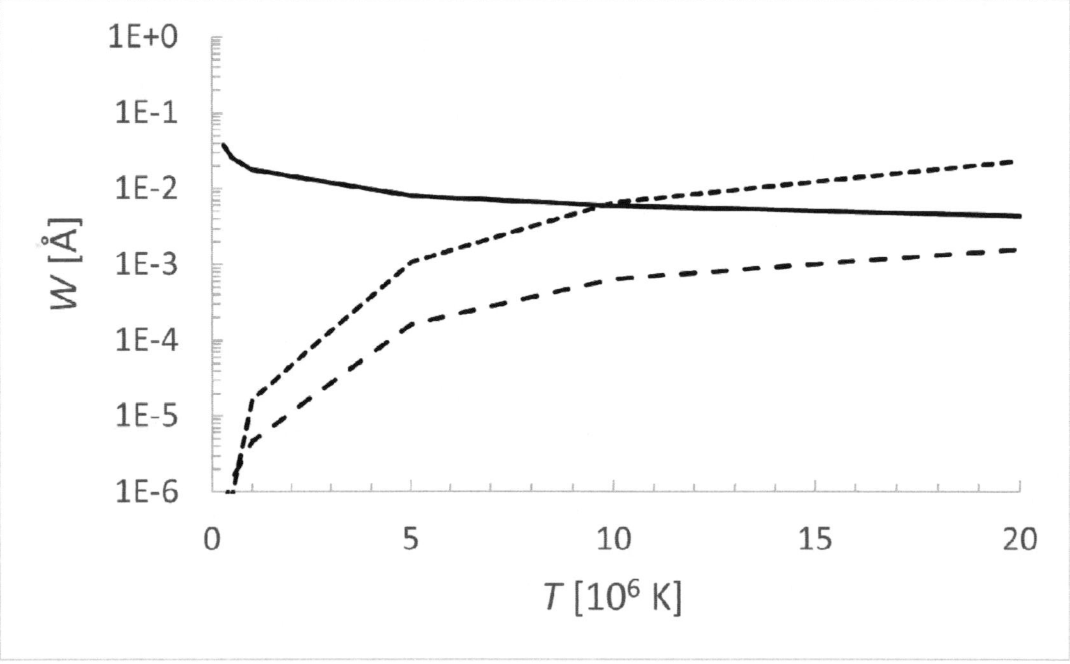

Figure 1: Behavior of electron-impact (solid line), proton-impact (lower dashed line) and Fe XXVII-impact (upper dashed line) width of Fe XXV $2s^1S$-$2p^1P^o$ spectral line with temperature. Perturber density is 10^{20} cm^{-3}.

In Fig. 1 is presented the behavior with temperature of Stark width for Fe XXV $2s^1S$-$2p^1P^o$ spectral line for broadening by electron-, proton-, and Fe XXV ion-impacts. We can see that with the increase of temperature ion broadening increases and becomes comparable with electron broadening in the case of proton broadening and dominant in the case of Fe XXV ion broadening, on temperatures characteristic for atmospheres of neutron stars.

The complete results of this study, as well as the corresponding analysis, are published in Dimitrijević et al. (2023). They are of interest for investigation and modelling of neutron star atmospheres and their environment and inertial fusion plasma. Also, we intend to implement obtained results for Stark widths and shifts of Fe XXV spectral lines in the STARK-B database (Sahal-Bréchot 2010, Sahal-

Bréchot et al. 2012, 2015, 2022), a part of Virtual Atomic and Molecular Data Center VAMDC (Dubernet et al. 2010, 2016, Albert et al. 2020).

Acknowledgments

This work has been supported with a STSM visit grant CA16214-48065 for M.S.D. within the framework of COST Action CA16214 "The multi-messenger physics and astrophysics of neutron stars, PHAROS".

Partial support from Faculty of Applied Mathematics and Informatics, Technical University of Sofia, Bulgaria is also acknowledged. This work has been partially supported and by the Paris Observatory, the CNRS, and the PNPS (Programme National de Physique Stellaire, INSU-CNRS), France.

References

Albert D., Antony B. K., Ba Y. A., Babikov Y. L., Bollard P., Boudon V., Delahaye F., Del Zanna G., Dimitrijević M. S., et al.: 2020, *Atoms*, 8, 76.

Barnard A. J., Cooper J., Smith E. W.: 1974, *J. Quant. Spectrosc. Radiat. Transfer*, 14, 1025.

Bates D. R., Damgaard A.: 1949, *Philos. Trans. R. Soc. London A*, 242, 101.

Cottam J., Paerels F., Mendez M.: 2002, *Nature*, 420, 51.

Cowley, C. R.: 1971, *Observatory*, 91, 139.

Dimitrijević M. S., Christova M. D., Yubero C., Sahal-Bréchot S.: 2023, *MNRAS*, 518, 2671.

Dimitrijević M. S., Sahal-Bréchot S.: 1984, *J. Quant. Spectrosc. Radiat. Transfer*, 31, 301.

Dimitrijević M. S., Sahal-Bréchot S.: 1996, *Phys. Scr.*, 54, 50.

Dimitrijević M. S., Sahal-Bréchot S.: 2014, *Atoms*, 2, 357.

Dimitrijević M. S., Sahal-Bréchot S., Bommier V.: 1991, A&AS, 89, 581.

Dubernet M. L., Boudon V., Culhane J. L., Dimitrijević M. S., Fazliev A. Z. et al.: 2010, *J. Quant. Spectrosc. Radiat. Transfer*, 111, 2151, http://www.vamdc.org

Dubernet M. L., Antony B. K., Ba Y. A., Babikov Yu. L., Bartschat K., et al.: 2016, *J. Phys. B*, 49, 074003.

Fleurier C., Sahal-Bréchot S., Chapelle J.: 1977, *J. Quant. Spectrosc. Radiat. Transfer*, 17, 595.

Griem H. R.: 1974, *Spectral line Broadening by Plasmas,* McGraw-Hill, New York.

Kramida A., Ralchenko Yu., Reader J., and NIST ASD Team: 2021, NIST Atomic Spectra Database (ver. 5.9), [Online]. Available: https://physics.nist.gov/asd [2022, November 7], National Institute of Standards and Technology, Gaithersburg, MD.

Madej J.: 1989, *A&A*, 209, 226.

Majczyna A., Madej J., Joss P. C., Rózanska A.: 2005, *A&A*, 430, 643.

Moon S. J., Wilks S. C., Klein R. I., Remington B. A., Ryutov D. D., et al.: 2005, *Astrophysics and Space Science*, 298, 293.

Oertel G. K., Shomo L. P.: 1968, *ApJS*, 16, 175.

Paerels F., 1997, *ApJ*, 476, L47.

Sahal-Bréchot S.: 1969a, *A&A*, 1, 91.

Sahal-Bréchot S.: 1969b, *A&A*, 2, 322.

Sahal-Bréchot S.: 1974, *A&A,* 35, 319.
Sahal-Bréchot S.: 1991, *A&A*, 245, 322.
Sahal-Bréchot S.: 2010, *J. Phys.: Conf. Ser.*, 257, 012028.
Sahal-Bréchot S.: 2021, *Atoms*, 9, 29.
Sahal-Bréchot S., Dimitrijević M. S., Ben Nessib N.: 2014, *Atoms*, 2, 225.
Sahal-Bréchot S., Dimitrijević M. S., Moreau N.: 2012, *J. Phys.: Conf. Ser.*, 397, 012019.
Sahal-Bréchot S., Dimitrijević M. S., Moreau N., Ben Nessib N.: 2015, *Phys. Scripta*, 50, 054008.
Sahal-Bréchot S., Dimitrijević M. S., Moreau N.: 2022, *STARK-B database,* [online]. Available: http://stark-b.obspm.fr [November 1, 2022]. Observatory of Paris, LERMA and Astronomical Observatory of Belgrade.
Shirai T., Sugar J., Musgrove A., Wiese W. L.: 2000, *J. Phys. Chem. Ref. Data*, Monograph No. 8, AIP Press, Melville, NY.
Sugar, J., Corliss, C.: 1985, *J. Phys. Chem. Ref. Data* 14, Suppl. 2.
Suleimanov V. F., Klochkov D., Pavlov G. G., Werner K.: 2014, *ApJS*, 210, 13.
Van Peet J. C. A., Costantini E., Méndez M., Paerels F. B. S., Cottam J.: 2009, *A&A,* 497, 805.
Van Regemorter H., Hoang Binh Dy, Prud'homme M., 1979, *J. Phys. B*, 12, 1073.
Werner K., Nagel T., Rauch T., Suleimanov V.: 2007, *Adv. Space Res.*, 40, 1512.

Proceedings of the XIII Bulgarian-Serbian Astronomical Conference (XIII BSAC) Velingrad, Bulgaria, October 3-7, 2022
Editors: Evgeni Semkov, Milan S. Dimitrijević, Momchil Dechev and Zoran Simić
Publ. Astron. Soc. "Rudjer Bošković" No 25, 2023, 63-67

INTERGRAM INSTEAD HISTOGRAM: PROGNOSIS OF THE EXTREME POPULATION VALUE

TSVETAN B. GEORGIEV

Institute of Astronomy and NAO, Bulgarian Academy of Sciences,
72 Tsarigradsko Shaussee Blvd, Sofia 1784, Bulgaria
E-mail: tsgeorg@astro.bas.bg

Abstract. The intergram is a proxy of the histogram. In contrast to the histogram and similar to the cumulative distribution, the intergram is unique for the studied data sample. The intergram is built by the method of the shortest intervals containing specified part among the sorted data. The extrapolation of the bright leg of interrgram in log-log coordinates may prognosticate the extreme value of the studied sample for 1% of the population. Four examples of extrapolation and comparison between the prognosis and observation are shown. The distribution of the diameters and masses of 76 bodies of the Solar system give evidences about one missing big planet. The estimated maximal rotation velocities among of 639 disk galaxies, 242 km/s, is in good agreement with the observations. The habitant numbers of 297 Bulgarian towns shows absent of town with about 525 000 habitants. The excess is the metropolis, containing 2.5 times more habitahts. Also, the brightest stars of the cluster Praesepe pose apparently about 6.2 mag. However the intergram of 196 cluster members prognosticates shows that in the beginning 1% of the stars had to had about 4 mag. About 5.6% of the sample is already out of the MS.

1. INTRODUCTION

Many astronomical values show asymmetric distributions, sometimes with triangular shape. Such are the Wolf number of the solar spots, the light fluxes of flickering stars (Georgiev et al. 2022). Such are also the values of some physical parameters – masses and diameters of planetary bodies, luminosities and diameters of galaxies, etc. These asymmetric distributions seem to be so fundamental in the nature like the triangles in the geometry. That's why we proposed a special proxy of the histogram, called intergram, with 21 applications (Georgiev 2022).

Figure 1 explain the building of intergram of the distribution of 76 bodies of the Solar System by their log-diameters. These bodies (planets, satellites, asteroids) are selected to be larger than 180 km. The Sun and 4 giant planets are excluded. The intergram is shown by its two appearances in Figs.1(a) and 1(b).

In practice, the scientist does not know the density function of the studied population. He has only a "good sample" of *n* data, sorted by increasing. Such data are shown by dots along the ordinate in Fig. 1(a) and along the abscissa in Fig. 1(b). A relevant histogram, that represents approximately the density function of this sample is shown in Fig. 1(b) by dashed curve. The appearance of the histogram depends on the step and binning interval, specified by the user. Another representation of the sample density, the inerGram, is possible through the method of the Minimal Interval of Data, containing *p*% of the sorted data (*p*%-MID). This method is mentioned by Rousseeuw & Leroy (1987) in a connection with their "method of the least trimmed squares". Here we draw the shape of the SP through the bounds of the *p*%-MIDs.

2. BUILDING THE INTERGRAM

Figures 1 showS details in the building of the intergram.

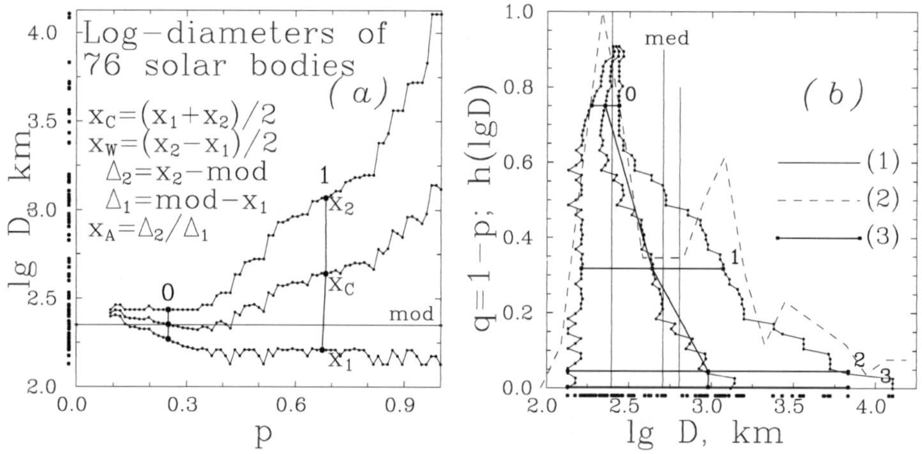

Figure 1: Intergram for solar bodies. (a) Definition form of the intergram tripod $f(\lg D_k)=f(p)$ with localparameters: x_1 and x_2 – bounds, x_C – center, x_W – width, x_A – asymmetry; (b) Working form of the intergram $q_k=f(\lg D)$ where $q=1-P$, D is the diameter in the sorted sample. (1) – intergram; (2) – histogram; (3) – intergram skeleton (see Georgiev 2022).

Let us to have a sorted data sample $\{X\}$: $x_1, x_2, ..., x_n$. Preliminary <u>Step 0</u> is a choice of the initial minimal part of data, p_1, for omitting the top of the inergram. Often the modal part of the sample dencity is complicate and p_1 must be enough large. In Fig. 1 p_1=0.1 (6 data). A better value may be p_1=0.25 (Figs. 2, 3). In Fig.4 p_1=0.5. <u>Step 1</u> is the deriving of the 1st minimal interval $[x_1, x_2]$ among the sorted data, containing $m_1=p_1 \times n$ data. <u>Step 2</u> is deriving the 2nd minimal interval, containing $m_2=m_1+1$, data with relevant part of data $p_2=m_2/n$. <u>Step 3</u> concerns by the same way $p_3=m_2+1$ etc., till $m_{k-1}=n-1$. In the end the case $m_k=n$ may be added for $p_k=1$. The number of the steps (of used MIDs) is $n-m_1+1$.

The result in Fig.1(a) is the intergram by definition. It is a "lie" tripod graph with abscissa p and 3 ordinates – MID bound points x_1 and x_2, as well as their average value x_C. Here 70 abscissa points are used.

Figure 1(a) contains two vertical segments which showthe MIDs above the abscissae p=0.25 (0) and p=0.6826 (1). The middle point of the 25%-MID is used as a robust mode estimator for the syudied distribution. The 68.26%-MID corresponds to the part p for 1 sigma interval of the normal distribution. The vertical segments for 2 MIDs in Fig. 1(a) are a part of the interval skeleton of the sample. The full skeleton contains also segments for 2 sigma and 3 sigma intervals of the samle in Fig. 1(b). The skeleton gives morphological parameters of the distribution (see Georgiev 2022).

It is convenient to represent the standing intergram tripod, as in Fig. 1(b), simi-lar to histogram. For this purpose we change p with $q=1 - p$ and yye use q as ordinate. In Fig.1(b) the dashed curve is the histogram. The verticals show the positions of the mode, median and average.

In principle the distribution with positive skewnes has exponentially decreasing right leg. Under a log-ordinate this leg becomes quasi-linear. Thus the relevant fit may prognosticate the quantity of the population for given abscissa value, for example for 1% of the population (see below).

3. APPLICATIONS OF THE INTERGRAM PROGNOSIS

Figures 2 show intergrams of Solar System bodies larger than 180 km.

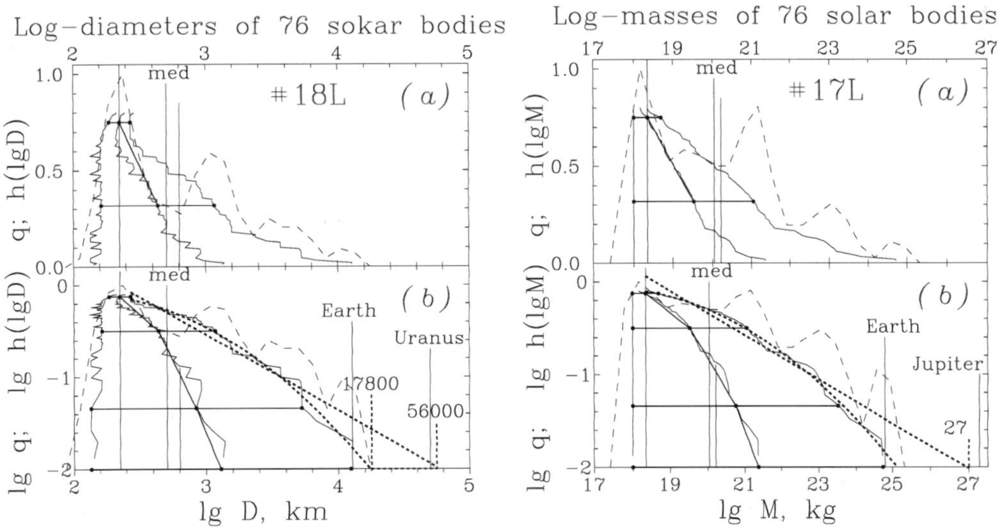

Figure 2: Intergrams of the solar bodies by diameters (left) and masses (right). Thin curves and lines show the integrams and their skeletons, dashed curves show histograms, short dashed curves show linear and quadratic fits of the intrgrams in log-log coordinates.

The relevant histograms (dashed curves) show excess of bodies with sizes about 1000 km and masses about 10^{21} kg. Such bodies are similar to the satelites of Ura-nus. The intergrams do not show suc details. The right linear extrapolations of the down intergrams (in the left without Sun and giant planets, in the right without Sun) prognodtiecte that 1% of the population must contains one big planet arge as Uranus and as massive as Jupiter, respectively. The intergram of the densities predict a planet with mean density about 8.6 g/cm^3 Georgiev (2022).

Figure 3, left diagrams, shows intetgrams of the the rotational velocities of 639 disk galaxies. This example concerns a distribution, depending on numerous na-tural factors. In (b) the right leg sows linearity from 10 to 200 km/s. The prog-nostic value for 1% MID is V_R=242 km/s. This prognosis is in good agreement with the observations. Only three more massive galaxies are present in the sample, including 1 with V_R =277 km/s.

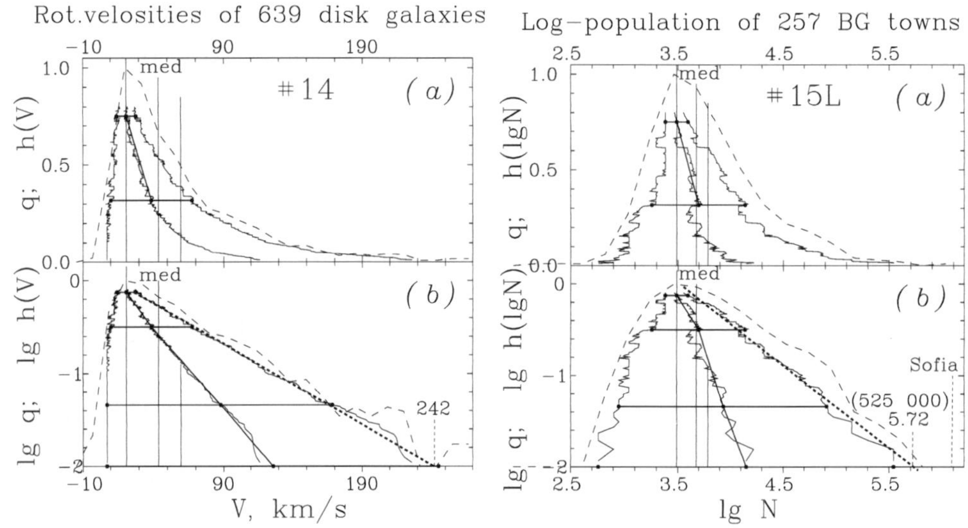

Figure 3: Left diagrams: Intergrams of the rotation velocities of disk galaxies; Right diagrams: Intergrams of the habitants of Bulgarian towns. Thin curves and lines – integrams and skeletons; Dashed curves – histograms; Short dashed curves – linear fits and extrapolations of the bright legs of the intrgrams in log-log coordinates.

Figure 3, right diagrams, shows the intergrams of the habitants of 257 Bulga-rian towns. This is not natural distribution It depends mainly on the human factors. The prognostic value for 1% MID must be 525 000, more thah Plovdiv or Varna, which have about 330 000. However, the metropolis with 1 243 000 habitants.is about 2.5 times larger in respect with the prognosis.

Figure 4 shows an application of the intergram for prognosis of the magnitude of the brightest stars in the beginning time of the open cluster Praesepe (M 44). In the catalogue of Wang et al. (1995) 196 stars are members of the cluster. Their CMD is represented in Fig. 4(a). V-intergram of these stars is shown in Frig.4(b).

Thh3 intrgram in Fig. 4(b) is natural sample distribution, depending on the evolutionary status of the cluster. A part of the brightest stars are leaved the main sequence. The prognostic value for 1% MID, given by the linear fit is about 4 mag. The gradient of the fit is -0.276. The breakdown level of the fit is lg q(V) = -1.25,, i.e. 5.6% of the studied sample sre leaved the MS of the CMD and 94.4% remain. The use of B-magnitudes gives the same result, 4 mag but the gradient is 0.259.

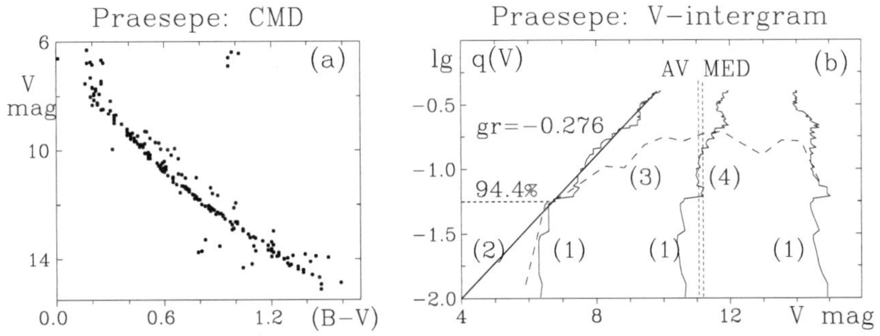

Figure 4: (a) Apparent color-magnitude diagram of 196 members of the cluster Praesepe; (b) Log-intergram of the V-mag-nitudes: 1 – legs of the intergram; (2) fit and extrapolation of the bright leg; (3) histogram; (4) –average abd median of the distribution.

4. CONCLUSIONS

The intergram as a procy of the histogram is an useful unique represent-tation of the data distribution. The parameters implemented in that are implement-ted in Fig. 1(a) may be used for morphological functions of the intergram – beha-viors of the center $C(p)$, width $W(p)$ and asymmetry $A(p)$. The gradients of these func-tions give classifications of the intergrams (see for details Georgiev, 2022).

In this work the intergram is used as tool for prognosticate the extreme values in the studied population.

Acknowledgements

The author is grateful to O. Stanchev and A. Mutafov for the attention and help.

References

Georgiev Ts.: 2022, *Bulg. Astron. J.*, 36, 70-81.

Georgiev,Ts., Boeva S., Latev G., Semkov E., Stoyanov, K. A., Tsvetkova S.: 2021, *Bulg. Astron. J.*, 34, 10-29.

Rousseeuw P. J., Leroy A. M.: 1987, *Robust Regression and Outlier Detection,* John Willy & Sons.

Wang J.J., Chen L., Zhao J.H., Jiang P.F.:1995, *Astron. Astrophys. Suppl.*, 113, 419-430.

Proceedings of the XIII Bulgarian-Serbian Astronomical Conference
(XIII BSAC) Velingrad, Bulgaria, October 3-7, 2022
Editors: Evgeni Semkov, Milan S. Dimitrijević, Momchil Dechev and Zoran Simić
Publ. Astron. Soc. "Rudjer Bošković" No 25, 2023, 69-74

REVEALING OF QUASI-PERIODS IN TIME SERIES BY STRUCTURE LENGTH EMINENCE

TSVETAN B. GEORGIEV

Institute of Astronomy and NAO, Bulgarian Academy of Sciences,
72 Tsarigradsko Shaussee Blvd, Sofia 1784, Bulgaria
E-mail: tsgeorg@astro.bas.bg

Abstract. We present conceptually simple method for revealing of quasi-periods in time series. The method is based on a function of the structure length eminence, whose maxima positions correspond to quasi-periods (or real periods). The method may be regarded as an analogue of the periodogram analysis, but it works in the real domain dealing with spectrum of periods. The method produces average profile of any specified repeating structure. The method is applicable on time series with constant point step and neglect large scale trend. Otherwise, the time series may be resampled and the large scale trend may be followed and removed by local smoothing. One test of the method is perform on the flickering of the symbiotic star T CrB. System of quasi-periods is revealed.. The basic periods 9.9 min ±8% and 16.3 min. Other test is performed on the prognostic behavior of the atmosphere ^{14}C down to 15 200 years BC. Two basic periods, of 950 yr ± 8% and 2235 yr ± 3% are revealed.

1. INTRODUCTION

Recently we revealed quasi-periods (QPs) in time series (TSs) of the flickering active symbiotic stars RS Oph, T CrB, and MWP 560 (Georgiev et al., 2022). In such cases the frequency methods are difficult to apply, because the TS is short, containing irregular large scale trend and strong noise. We removed preliminary the large scale trend globally, by a polynomial fit over the whole TS. Then we detected QPs by the minima positions of the structure function (SF). Later we resampled the TS within a constant step and detected the same QPs by the maxima positions of the autocorrelation function (ACF). However, these methods derive QPs with a low accuracy and a low resolution, omitting short QPs. By this reason we proposed a method of the structure length eminence (MSLE) (Georgiev, 2023). It checks numerous pixel lengths *m* as lengths of possible repeating structures and build a function of the structure length eminence (SLE), E(*m*) , (Eq. 1, 2, Fig.~1e).

Really E(m) is a spectrum of the QPs. Any maximum position of the SLE is a QP, but the real periods may be distinct additionally.

The goals of the present paper are to describe the MSLE and to show two applications. We regard one usual example of star flickering and one complicate example about the prognostic behavior of the ^{14}C in the past. Other examples are discussed in Georgiev (2023).

The used abbreviations follow:
ACF – autocorrelation function;
AV – average value;
MSLE – Method of the structure length eminence;
TS – time series;
QP– quasi-period;
SD – standard deviation;
SF – structure function;
SLE – structure length eminence;
SS – significant (repeating) structure;
SW – smoothing window size.

2. METHOD OF THE STRUCTURE LENGTH EMINENCE

Suppose to have a TS with n equally spaced datapoints ($F(t_i)$ (i=1,n), with zero global average and neglect large scale trend. Such a TS is the residual TS (Fig.1b), derived from the input TS (Fig. 1a) by a large scale trend removal. Suppose that the TS contains a significant (repeating) structure (SS) with an unknown length of m_S pix. We select m_S checking numerous data lengths m_1 m_2, ... , m_S, ..., m_M (e.g. 20--200 points), as follows.

Beginning with m_1 we pull up the first m_1 values of the TS into a (initially empty) set with cell numbers j=1,m_1. Then, we add there the next m_1 values from the TS. Later we add the next m_1, etc. In such a way we perform $k_1=n/m_1$ (integer number) additions. Then for every $_j$-th cell, containing k_1 additions, we derive the average values (AV) $a(j,m_1)$ and standard deviations (SD) $d(j,m_1)$. The program performs this procedure for every m between m_1 and m_M, deriving the relevant m-th profiles of the structure in AVs and SDs.

Latter the program derives the average amplitude $A(m)$ and the average SD $D(m)$ for every m-th profile:

(1) $$A(m)=<|a_j|>_m, \qquad D(m)=<d_j>_m.$$

Note that $A(m)$ gathers the absolute AVs $|a_j|$ and it is the average amplitude of the structure with the length m pix. Also, $D(m)$ is the average noise in this structure.

When a SS with length of m points is present, it produces maximum of $A(m)$ and minimum of D(m) (Georgiev, 2023, Fig. 1e). Therefore, a more sensitive indicator of SSs is the ratio $A(m)/D(m)$. For convenience, we use it multiplied by 100 and call it *structure length eminence*, SLE:

(2) $$E(m)=100 \times A(m)/D(m).$$

Each value of the SLE, $E(m)$, may be regarded as a kind of a signal-to-noise ratio, expressed by percentages in respect with the noise. Really, the local noise continuum of the SLE must be accounted for a correct estimation of the signal-to-noise ratio. The MSLE requires constant pixel step. In many cases the TS may be resampled by a linear interpolation.

The MSLS is applicable on TS without significant large scale trend and with zero global AV. Therefore, a preliminary amplitude decomposition of the input TS F_I into smoothed TS F_S and residual TS, $F_R=F_S - F_I$, is necessary. The subject of the MSLE is the residual TS $\Delta F(t)$, expressed in percentages of the smoothed TS:

(3) $$\Delta F(t) = 100 \times [F_I(t) - F_S(t)] / F_S(t).$$

Using such a residual TS the results of various MSLE applications may be compared easy.

The simple MSLF is applicable when the large scale trend may be fitted and removed by a low degree polynomial (Georgiev, 2023, Fig. 1). Usually the large scale trend is complicate and the polynomial is not enough flexible. *The basic MSLF* is applicable when the large scale trend is complicate, but it may be followed and removed by a sliding average (Figs. 1 and 2 here). The practice shows that the SW must be about 1.5 times larger than the supposed basic period.

3. FLICKERING PERIODS OF T CRB

We studied the flickering of the symbiotic star T CrB in U-band by the methods of the ACF and SF (Georgiev et al. 2021). Figure 1 represents the application of the MSLE on its light curve *u*12. Here the original TS with step of about 1/6 min is resampled to have a step of 1/10 min everywhere and then it is used as an input TS. Such transform performs smoothing of the jagged TS, and decreases the slope of its SF, but it does not influence the period estimation and ensure better describing of the short period profile shapes.

Figure 1(a) shows the input TS in U-band fluxes and the smoothed TS, both in dependence on the monitoring time t_M. Here AV and SD relate with the input TS. N is the number of the data points. Thee input TS contains complicate trend and strong noise.

Figure 1(b) shows the residual TS (Eq. 3), in percentages toward the smooth level, with zero AV and relevant SD. T_M is the monitoring time. SW = 161 pomys = 16.1 min is the window of the average smoothing, shown by the segment W. The edges of the TS are lost and the length of the residual TS is already 818 point. The residual TS in (b) shows 5 humps and a hint of about 17 min period.

Figures 1(c) and 1(d) show the ACF and the SF of the residual TS in (b), with respective steps τ and θ. These functions detect the shortest period p of 9 – 10 min and a systems of QPs like *p*, 2*p*, 3*p,* etc., close in the frames of ±10%. The half width ay the half of the humps is about 20%. The basic periods are not clear.

Figures 1(e) shows the SLE (Eq. 2) as a function of the structure time length t_L of the checked structure (with relevant data length *m*). The range is 4 – 34 min or 40 – 340 pix. A few humps protrude above the continuum of the SLE. Two of them (dots), at $P_1 = 9.9$ min and $P = 16.3$ min, seem to be the basic ones.

The revealing of the basic period includes an account of the shapes of the profiles. (See for details Georgiev, 2023, Figs. 1 and 2): (I) These periods have the highest maximums (above the SLE continuum), having also single profile shapes in (g) and (h); (II) They have clear counterparts at $2P_1$, $3P_1$ and $2P$ (asterisks), with double profiles shapes, like $2P_1$ in (i); (III) The basic period *P* has the expected "harmonic" counterparts at *P*/2, 3*P*/4 and 2*P*/3 (circles), again with single profiles, like *P*/2 in (f). These QPs are marked by circles; (IV) Here yhe period P_1 does not coincide with the harmonic at 2*P*/3, which is observed rarely, therefore is may be regarded as basic period too. By unknown reason its profile of P_1 (g) is far from a sinusoid shape.

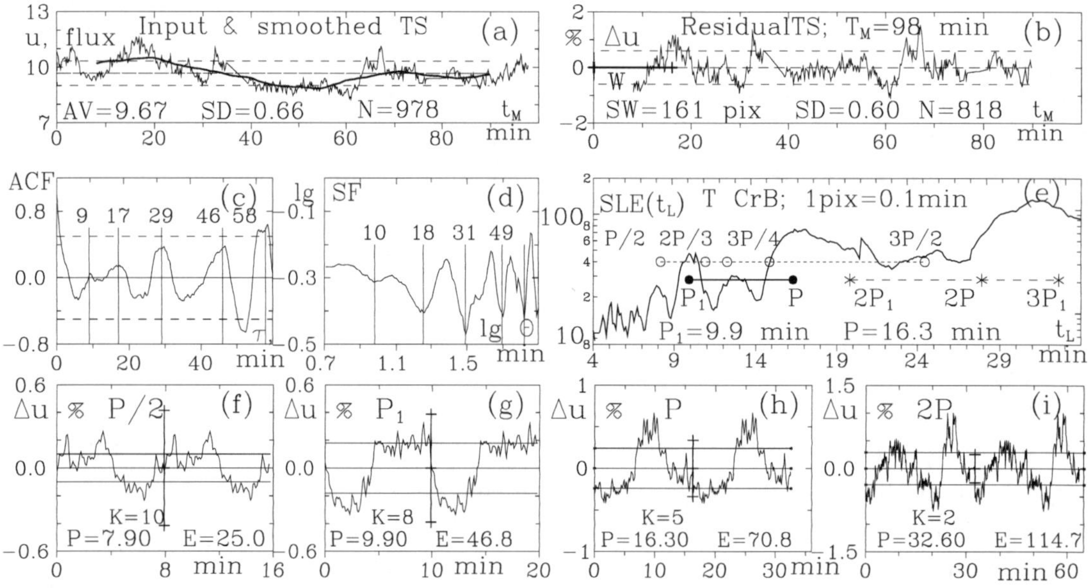

Figure 1: Application of the MSLE on the flickering of T CrB. (a): Input TS and Smoothed TS; (b): Residual TS; (c): ACF; (d): SF; (e): SLE – spectrum of the QPs where only P_1 and *P*are the basic periods; (f – i): Profiles of the SS.

Figures 1(f) – 1(i) show average profiles of SS, taken by the residual TS In (b). In the diagram *P* is the profile length. There *E* is the profile eminence, i.e. the absolute height of the relevant maximum in the SLE, without accounting of the SLE continuum. *P* corresponds to the highest *E*-value in the relevant hump in the SLE. *K* is the number of the additions that form the average profile.

So, unlike to the methods of ACF or SF, the MSLE recognizes the basic periods. The half width at the half of the maximum of the P_1-hump is 0.9 cen, i.e.

the external error is 8%. Here the *P*-hump in (e) is not single. Other examples are given in Georgiev (2023). There a comparing with the periodogram analysis of the flickering of AQ Men of Iłkiewich et al. (2021) is included.

4. PERIODS IN THE BEHAVIOUROF THE ^{14}C IN THE FAR PAST

Stuiver et al. (1998) have been published the restored behavior of the relative atmosphere ^{14}C concentration, up to 15 200 years BC. Here the original TS is preliminary resampled to have everywhere 1 step = 1 century and the input TS contains 1520 pix. The input TS (Figs.2(a)) is complicated and noised. The right tail of the TS seems corrupted and only 1325 points are processed by the MSLE.

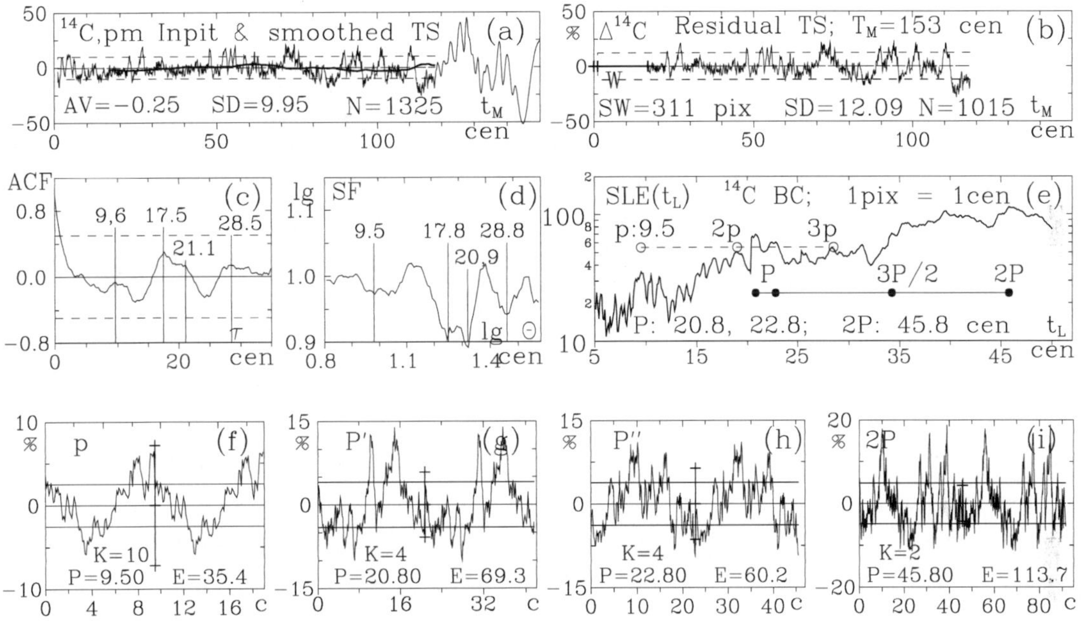

Figure 2: Application of the MSLE on the restored concentration of ^{14}C in the far past. (a): Input TS and smoothed TS in per mille; (b): Residual TS; (c, d): ACF and SF; (e): SLE (the spectrum of the periods); (f – i): Profiles of SS.

Figure 2 illustrates another application of the MSLE, like to the case in Fig. 1. While an eminent period of 23 cen. is suspected to be the basic one, the input TS in (a) is decomposed into smoothed TS and residual TS in (b) by SW size of 311 points. The edges of the TS are lost and the length of the residual TS is already 1015 points. Figures 2(c) and 2(d) show the ACF and the SF. They detect similar systems of QPs, but with low resolution. The basic QP is not clear.

The SLE in Fig. 2(e) is very complicate. The shortest significant period is *p*=9.5 cen. ±8%. The relevant average profile is given in (f). The basic period *P* appears at 20.8 and 22.9 cen., as well as 2*P* at 45.8 cen. The first two periods show single profiles in (g) and (h), respectively. The 2*P* period shows the expected double profile in (i). Obviously the ACF and SF in (c) and (d) recognize

$2p + P$ together. The use of the positions of the relevant humps in the MSLE, including $2P$ with double weight, gives the period of the basic cycle to be 22.35 cen. ±3%.

Generally, the Hallstatt Sun cycle of 2200 – 2400 years (Usoskin et al. 2016; Komitov & Kaftan 2003) obviously influences the concentration of the ^{14}C in the atmosphere. The use of the positions of the relevant humps in the MSLE gives the period of the Hallstatt cycle to be 2235 yr.

5. CONCLUSIONS

The MSLE is an analogue of the periodogram analysis, but it works in the real domain. The MSLE is useful and illustrative tool for revealing of QPs and real periods in TS with irregular trend. Four other examples are given in Georgiev (2023).

The SLE is built over the residual TS, with zero global average (Figs. (b)). The MSLE reveals the QPs with high resolution (Figs. (e)) and produces useful profiles of the repeating structures.

The MSLE is applicable when the large scale trend is the obstacle for the period deriving. Then the trend is following and removing by a sliding average with SW size about 1.5 times larger than the supposed large basic period.

Acknowledgements

The author is grateful to Krystian Iłkiewicz and to Rumen Bogdanovskifor for the attention and recommendations.

References

Georgiev Ts.B.: 2023, *Bulg. Astron. J.*, 38 (in print).

Georgiev Ts.B., Boeva S., Stoyanov K., Latev G., Spassov B., Kurtenkov A.: 2022, *Bulg. Astron. J.*, 37, 62-75. (gbs22)

Georgiev Ts.B., Boeva S., Latev G., Semkov E., Stoyanov K.A, Tsvetkova S. V.: 2021, *Bulg. Astron. J.*, 35, 10-29.

Iłkiewicz K., Scaringi S., Court J.M.C., et al.: 2021, *MNRAS,* 503, 4050-4060.

Komitov B., Kaftan V.: 2003, *Internat. J. of geomagnetism and aeronomy*, 43/5, 553.

Stuiver M., Reimer P.J., Bard E. et al.: 1998, *Radiocarbon,* 40, 1041-1083.

Usoskin I,G., Gallet Y., Lopes F., et al.: 2016, *Astron. Astrophys.*, 587, A150.

Proceedings of the XIII Bulgarian-Serbian Astronomical Conference (XIII BSAC) Velingrad, Bulgaria, October 3-7, 2022
Editors: Evgeni Semkov, Milan S. Dimitrijević, Momchil Dechev and Zoran Simić
Publ. Astron. Soc. "Rudjer Bošković" No 25, 2023, 75-84

COMPARISON AND CONTROL STARS AROUND QUASARS SUITABLE FOR THE ICRF – GAIA CRF LINK

MILJANA D. JOVANOVIĆ[1], GORAN DAMLJANOVIĆ[1] and FRANÇOIS TARIS[2]

[1]*Astronomical observatory, Volgina 7, 11060 Belgrade, Serbia*
[2]*Observatoire de Paris - SYRTE, 61 av. de l'Observatoire, 75014 Paris, France*
E-mail: miljana@aob.rs, gdamljanovic@aob.rs, francois.taris@obspm.fr

Abstract. To link International and Gaia Celestial Reference Frame is necessary to observe and monitor set of sources visible in radio and optical wavelengths. From 2013 until 2019, we observed 47 active galactic nuclei (AGNs) which are candidate sources for the link between the frames. Our observations, in optical V and R bands were performed using eight telescopes from Serbia, Spain, Bulgaria, and Austria. The brightness of the sources and control stars were determined by differential photometry using suitable comparison stars. The obtained light curves of sources are significant for the understanding of the physical processes inside them. We tested the brightness in V, and R bands, and V-R colour (of sources, and their comparison, and control stars) with two statistical tests (Abbe's criterion, and F test), and some results are presented here.

1. INTRODUCTION

The third data release of Gaia mission - Gaia DR3 (Gaia Collaboration, Vallenari, A. et al. (2022)) is made public available, since 13 June 2022, and it is based on data collected between 25 July 2014, and 28 May 2017. The reference epoch for Gaia DR3 is 2016.0. The set of Gaia Early Data Release 3 is complemented with: object classifications for about 1.8 billion sources, astrophysical parameters for ~5.5 million objects, radial velocities for ~33 million stars, variability analysis, together with the epoch photometry for ~10.5 million sources, etc. The Gaia celestial reference frame 3 is based on the observations of quasi-stellar objects (QSOs) at optical wavelength. It could be linked with the International Celestial Reference Frame 3 (Charlot et al. (2020)) using a set of QSOs visible in the optical and radio domains. In paper Bourda et al. (2011) 47 active galactic nuclei (AGNs) were suggested for that link. We observed those AGNs from 2013 to 2016, and tested their brightness (calculated using stars from their vicinity). The results for the five most observed sources were published in paper Jovanović (2019). We analysed the britghness of 7 more sources, and in this paper we present the results of investigation of brightness variability of their comparison, and control stars. Three sources are flat spectrum radio quasars - FSRQs (0049+003, 1212+467, and 1612+378), three are BL Lacertae - BL Lac (0907+336, 1034+574, and 1242+574), and one (1429+249) has dual nature of FSRQ, and BL Lac. The sources redshifts *z* are: 0049+003 z=0.40, 0907+336 z=0.35, 1034+574 z=1.10, 1212+467 z=0.72, 1242+574 z=1.00, 1429+249 z=0.41, and 1612+378 z=1.53. Also, in this paper are given finding charts for all sources and comparison and control stars.

2. OBSERVATIONS

The observations were carried out with eight telescopes, two at Astronomical Station Vidojevica (ASV) of Astronomical Observatory of Belgrade, Serbia; Joan Oró telescope (TJO) at the Montsec Astronomical Observatory, Catalonia, Spain; three at Rozhen, NAO, Bulgaria, one in Belogradchik, Bulgaria; and one telescope at Leopold Figl at Vienna, Austria. The details about these telescopes, their mirror aperture, mounted CCD cameras and optical filters are presented in paper Jovanović et al. 2021.

Two or more CCD images were obtained in *V* and *R* bands except when observations were done using TJO telescope, only one CCD image frame was taken in *V* and *R* bands. The image reduction was performed using advance image calibration in IRAF scripting language (ascl:9911.002)(Tody 1986, 1993). Bias, dark, and flat-field frames obtained for the same observing nights, are used for reduction, and for the bad pixel mapping (hot, and dead pixel map). L.A.Cosmic method by Pieter G. van Dokkum was used for the corrections for cosmic rays (van Dokkum, 2001).

We determined the brightness of the sources and control stars using differential photometry with Maxim DL software, with two or more comparison stars. The comparison and control stars were selected from the Sloan Digital Sky Survey Data Release 14 (SDSS DR14) catalogue (Abolfathi et al. 2018). Using the equations given by Chonis & Gaskell (2008) the SDSS PSF ugriz (point spread function u, g, r, i, and z) magnitudes were transformed into V, and R magnitudes

$$V = g - (0.587 \pm 0.022)(g - r) - (0.011 \pm 0.013),$$

$$R = r - (0.272 \pm 0.092)(r - i) - (0.159 \pm 0.022).$$

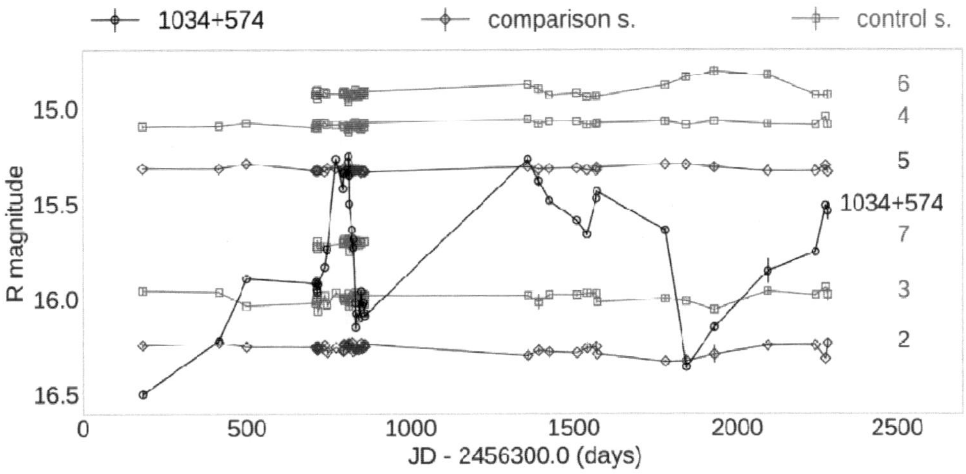

Figure 1: The light curves of 1034+574 (circles connected with lines), its comparison (diamonds connected with lines), and control stars (squares connected with lines).

The limits for *g, r*, and *i* magnitudes are 14.5 and 19.5 mag, for colour *r-i* are *0.08,* and *0.5* mag, and for colour *g-r* are 0.2, and 1.4 magnitudes. We chose comparison and control stars from the vicinity of the sources, the magnitudes of which satisfy these criteria and the colour is similar to the colour of the source.

In Table 1. are details of sources and their comparison (marked with *A* and *B*) and control stars: coordinates, the V_C and R_C magnitudes of stars (obtained using mentioned equations), V_O and R_O (average magnitudes from our observations), with number of observations N_{Vo} (in *V*), and N_{Ro} (in *R band)*. The input magnitudes for differential photometry (V_C and R_C magnitudes of stars) and calculated magnitudes V_O and R_O (from our observations) are within the error limits. The standard deviations of control stars have the same order of magnitudes as comparison ones.

An example of light curve, 1034+574 in R band during observational period, is presented in Fig. 1 with light curves of its comparison (number 2, and 5) and control stars (3, 4, 6, and 7). The light curve of source is marked with circles connected with lines, of its comparison stars with diamonds connected with lines, and of its control stars with squares connected with lines. The blazar brightness is variable unlike the brightness of comparison and control stars. The light curve of the control star with number 6 shows variability, and this is probably because the star was on edge of field of view (star with number 7 was not on each of the ccd images that were taken). Brightness of the blazar changed by more than 1 magnitude within this period. This blazar shows short-term variability, which is visible in TJO data.

The fields of view of 7 sources with its comparison and control stars are presented in Fig. 2. The source is marked with number 1, and the stars with remaining numbers. All fields of view are about 16'x16' obtained using 60 cm telescope at ASV with Apogee Alta U 42 CCD camera.

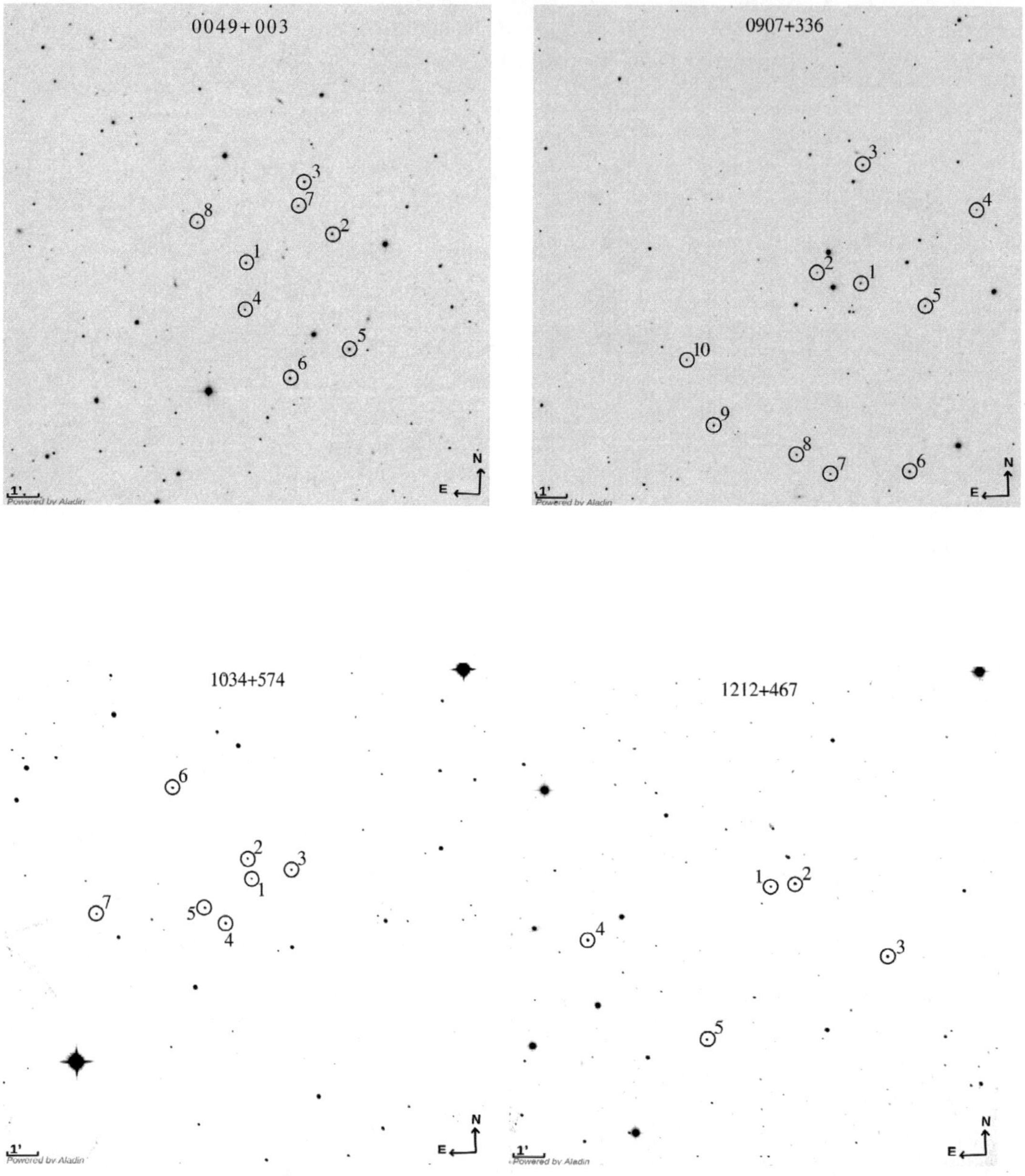

Figure 2: Fields of view of sources: 0049+003, 0907+336, 1034+574, 1212+467, 1242+574, 1429+249, and 1612+378

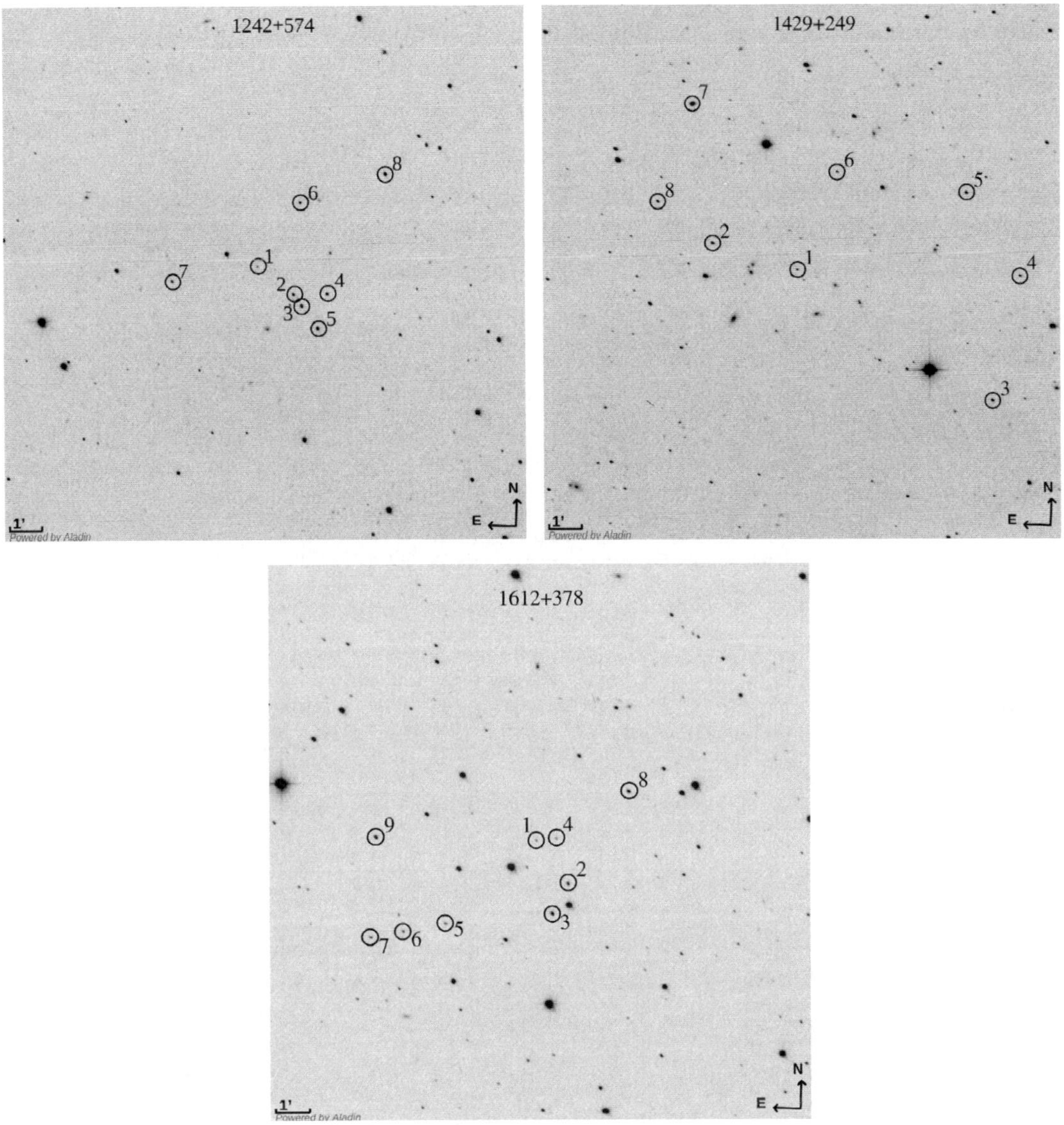

Figure 2: continued.

We compared our data to the data from Gaia Data release 3 (Gaia DR3). We calculated G_C for the stars with equation given by Satoretti et al. (2022) using our data in V, and R bands

$$Gc = V - 0.03088 - 0.04653(V - R) - 0.8794(V - R)^2 + 0.1733(V - R)^3, \sigma = 0.0352$$

and compare it with G magnitude given in Gaia DR3.

We presented in Fig 3. the data for the stars (comparison and control) of 12 sources (7 presented here, and 5 in paper Jovanović et al. (2021). Coefficients of linear fit (intercept=0.082, and slope=0.9960) were estimated using the weighted least square method. The Pearson linear correlation coefficient is r=0.9991, and null hypothesis probability is very close to 0. With the slope and Pearson's coefficient around 1, and calculated probability it is shown that there is a correlation between our calculated and G magnitudes from the DR3 catalogue.

Table 1: Coordinates, *V* and *R* magnitudes of comparison and control stars in the fields of sources

Object № star	$\alpha_{J2000.0}$ (°)	$\delta_{J2000.0}$ (°)	$V_C \pm \sigma_{V_C}$ (mag)	$R_C \pm \sigma_{R_C}$ (mag)	$V_O \pm \sigma_{V_O}$ (mag)	N_{V_O}	$R_O \pm \sigma_{R_O}$ (mag)	N_{R_O}
0049+003	13.02321	0.593930						
2^A	12.97558	0.60950	16.721 ± 0.039	15.830 ± 0.068	16.715 ±0.026	30	15.835 ±0.013	40
3^B	12.99098	0.63657	16.303 ± 0.036	15.680 ± 0.042	16.307 ±0.018	30	15.673 ±0.010	40
4	13.02369	0.56957	17.253 ± 0.030	16.859 ± 0.033	17.265 ±0.075	26	16.876 ±0.049	36
5	12.96617	0.54902	16.367 ± 0.038	15.547 ± 0.053	16.333 ±0.044	20	15.509 ±0.034	27
6	12.99846	0.53368	16.821 ± 0.039	15.914 ± 0.067	16.796 ±0.043	15	15.902 ±0.022	24
7	12.99423	0.62415	16.988 ± 0.026	16.655 ± 0.027	16.973 ±0.060	26	16.637 ±0.035	36
8	13.05000	0.61540	17.392 ± 0.034	16.804 ± 0.040	17.402 ±0.063	26	16.795 ±0.049	35
0907+336	137.65431	33.49012						
2^A	137.68218	33.49568	16.947 ± 0.027	16.493 ± 0.032	16.983 ±0.041	39	16.535 ±0.031	42
3^B	137.65315	33.55212	15.152 ± 0.025	14.765 ± 0.029	15.143 ±0.010	36	14.755 ±0.009	39
4	137.57933	33.52884	16.754 ± 0.023	16.402 ± 0.029	16.727 ±0.048	37	16.392 ±0.045	38
6	137.62254	33.39133	15.595 ± 0.036	14.787 ± 0.053	15.664 ±0.019	14	14.816 ±0.011	14
7	137.67337	33.39007	16.600 ± 0.031	15.964 ± 0.042	16.673 ±0.029	13	15.998 ±0.014	13
8	137.69512	33.40002	15.840 ± 0.024	15.596 ± 0.025	15.841 ±0.040	14	15.581 ±0.026	14
9	137.74861	33.41535	15.412 ± 0.028	14.910 ± 0.031	15.439 ±0.021	11	14.920 ±0.011	11
10	137.76623	33.44946	16.320 ± 0.028	15.817 ± 0.033	16.349 ±0.024	5	15.831 ±0.025	5
1034+574	159.43461	57.19878						
2^A	159.43831	57.20934	16.764 ± 0.028	16.252 ± 0.036	16.770 ±0.025	47	16.262 ±0.024	47
5^B	159.48269	57.18367	15.874 ± 0.029	15.329 ± 0.040	15.872 ±0.011	47	15.323 ±0.011	47
3	159.39357	57.20304	16.654 ± 0.032	15.993 ± 0.046	16.662 ±0.041	47	15.999 ±0.027	47
4	159.46188	57.17536	15.714 ± 0.031	15.103 ± 0.042	15.708 ±0.024	47	15.088 ±0.014	47
6	159.51361	57.24663	15.351 ± 0.027	14.904 ± 0.034	15.349 ±0.048	41	14.918 ±0.032	41
7	159.59133	57.18112	16.480 ± 0.038	15.688 ± 0.056	16.509 ±0.035	25	15.709 ±0.016	24
1212+467	183.79143	46.45420						
3^A	183.70101	46.41680	16.053 ± 0.028	15.760 ± 0.030	16.036 ±0.020	49	15.749 ±0.020	49
2^B	183.77226	46.45566	15.782 ± 0.029	15.445 ± 0.032	15.802 ±0.017	50	15.460 ±0.019	50
4	183.93530	46.42732	16.455 ± 0.033	16.089 ± 0.035	16.404 ±0.029	16	16.036 ±0.021	16
5	183.84232	46.37444	17.171 ± 0.031	16.715 ± 0.035	17.124 ±0.057	25	16.671 ±0.047	25
1242+574	191.29167	57.16510						
3^A	191.25047	57.14550	15.605 ± 0.036	15.123 ± 0.031	15.620 ±0.012	49	15.138 ±0.008	57
6^B	191.25146	57.19683	16.806 ± 0.034	16.428 ± 0.032	16.770 ±0.029	43	16.383 ±0.022	51
2	191.25798	57.15121	16.184 ± 0.035	15.773 ± 0.031	16.186 ±0.021	49	15.781 ±0.023	57
4	191.22685	57.15156	15.837 ± 0.034	15.462 ± 0.029	15.840 ±0.023	49	15.459 ±0.017	57
5	191.23555	57.13461	15.190 ± 0.031	14.790 ± 0.029	15.146 ±0.018	49	14.761 ±0.016	56
7	191.37149	57.15773	16.593 ± 0.039	16.227 ± 0.029	16.559 ±0.026	42	16.192 ±0.033	50
1429+249	217.85787	24.70575						
2^A	217.90576	24.71909	16.336 ± 0.034	15.778 ± 0.039	16.340 ±0.028	40	15.787 ±0.032	44
6^B	217.83619	24.75416	17.459 ± 0.032	17.019 ± 0.033	17.449 ±0.037	33	16.991 ±0.031	37
3	217.74829	24.64108	16.622 ± 0.033	16.102 ± 0.039	16.586 ±0.038	29	16.053 ±0.054	29
4	217.73247	24.70287	17.391 ± 0.028	17.042 ± 0.032	17.373 ±0.065	20	16.988 ±0.057	21
5	217.76278	24.74408	16.377 ± 0.032	15.999 ± 0.030	16.344 ±0.039	32	15.973 ±0.047	32
8	217.93664	24.73984	16.753 ± 0.031	16.378 ± 0.031	16.712 ±0.031	28	16.340 ±0.036	28
1612+378	243.69564	37.76869						
4^A	243.68317	37.76964	17.007 ± 0.032	16.489 ± 0.041	17.020 ±0.032	31	16.515 ±0.022	36
2^B	243.67568	37.74841	15.529 ± 0.028	15.225 ± 0.033	15.530 ±0.014	37	15.223 ±0.018	42
3	243.68553	37.73414	15.096 ± 0.029	14.739 ± 0.034	15.082 ±0.012	37	14.722 ±0.022	42
5	243.75125	37.72934	16.433 ± 0.029	16.070 ± 0.034	16.407 ±0.039	37	16.024 ±0.048	42
8	243.63855	37.79195	15.039 ± 0.030	14.627 ± 0.035	15.033 ±0.032	31	14.609 ±0.038	36

Notes. $^{(A)}$ Refers to comparison star A. $^{(B)}$ Refers to comparison star B.

In Gaia DR3 catalogue two stars are marked with variability flag, C_2 of 1722+119 and the star with number 5 of object 1242+574. The C_2 was marked as variable also in paper Doroshenko et al. (2014). With magnitude less than 13 mag it is the brightest star from our list. For star number 5 we did not find signs of variability, and the star is close to the fitting line (calculated value and given from catalogue are similar). For the stars which are distant from the fitting line, we have small number of observations.

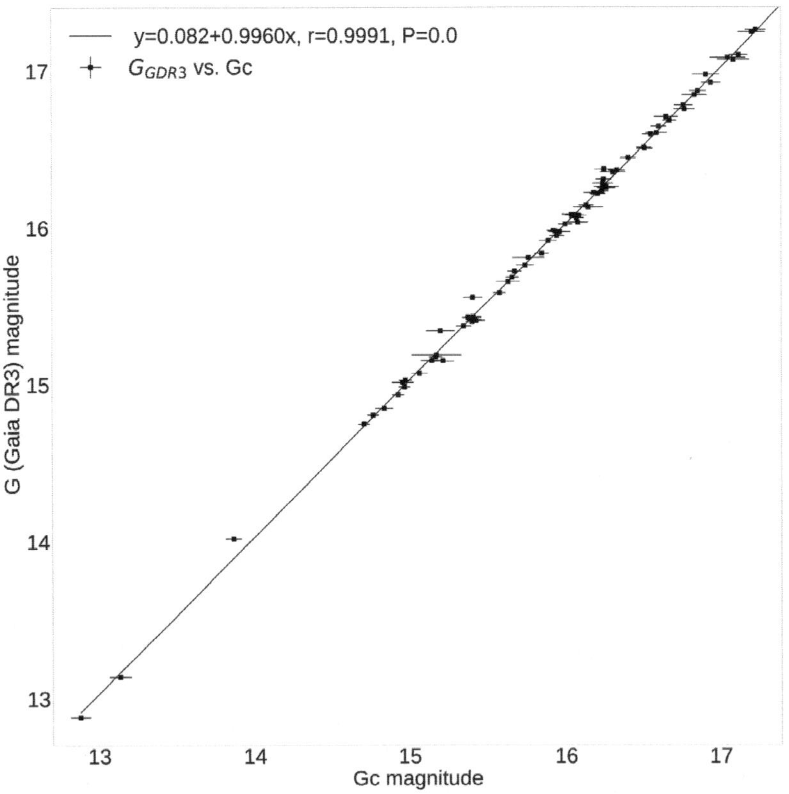

Figure 3: The linear fit of G magnitudes (Gaia DR3 vs calculated).

3. ANALYSIS METHODS AND RESULTS

To investigate variability in the control stars, we performed two statistics: Abbé's criterion and F-test. We consider that the stars are variable if the variability is detected by both tests. The tests require normal distribution of data, for that reason we applied 3-σ rule and Shapiro-Wilk test of normality (Razali et al. 2011). We discarded some of the data and concluded that the statistical methods can be applied.

Abbé's criterion is used for checking the absence of systematic changes in tested brightness. The statistic q is defined as the ratio of the Allan variance σ_{AV}, and unbiased sample variance σ_D,

$$q = \frac{\sigma_{AV}}{\sigma_D} = \frac{\frac{1}{2(n-1)}\sum_{i=1}^{n-1}(x_{i+1}-x_i)^2}{\frac{1}{n-1}\sum_{i=1}^{n}(x_i-\bar{x})^2} = \frac{1}{2}\frac{\sum_{i=1}^{n-1}(x_{i+1}-x_i)^2}{\sum_{i=1}^{n}(x_i-\bar{x})^2}$$

where $\bar{x}$ is the mean value of the magnitudes. The critical point q_C is

$$q_c = 1 + u_\alpha/\sqrt{n + 0.5(1+u_\alpha)^2}$$

where u_α is quantile of normal distribution for the significance level α. If q is lower than critical q_C, the elements of the sample cannot be accepted as random and independent. We calculate Abbé's statistics q_A, and q_B for two data sets: differences of magnitudes of control stars, and comparison stars A, and B, for the significance level α=0.001. Also we implement the Abbe statistic to the sample of brightness of comparison stars.

Also, we used F - test to determine brightness variability as in papers de Diego (2010), Jovanović (2019). The test statistic is

$$F = \frac{VarX}{VarY}.$$

The H_0: $Var\ X = Var\ Y$ is tested hypothesis, and H: $Var\ X > Var\ Y$ is alternative. We calculated three test statistics:

F_1 - where X is sample of differences of magnitudes of control stars and comparison star A, and Y sample of differences of magnitudes of control stars and comparison star B,

F_2 - where X is sample of differences of magnitudes of control stars and comparison star A, and Y is sample of differences of magnitudes of comparison star A and B,

F_3 - where X is sample of differences of magnitudes of control stars and comparison star B, and Y is sample of differences of magnitudes of comparison star A and B. The statistics F_1 is F_2 divided by F_3 and should be ~1, we expect that the brightness should be variable in the same manner (for both comparison stars A, and B). The three $F_{1,2,3}$ statistics are compared with the critical values F_c for he significance level 0.001, and number of freedom n-1, where n is the sample size. The hypothesis of non variability is discarded when F_1, and F_2 are greater than F_c.

The results of Abbé's criterion and F – statistics for control stars of sources are presented in Table 2. Four comparison stars have Abe statistics lower than the critical: star with number 3 of 1242+574 in V, and stars 3 of 0049+003, 3 0907+336, 2 1612+ 378 in R band. The variations of brightness of these stars are small, the standard deviation are 0.012, 0.010, 0.009, and 0.018, respectively.

Table 2: Statistical results of stars variability

object							
star	n	Abbé's criterion q_A, q_B, q_c V band	F–test F_1, F_2, F_3, F_c	star	n	Abbé's criterion q_A, q_B, q_c R band	F–test F_1, F_2, F_3, F_c
0049+003							
4	26	0.79 0.77 0.45	1.03 2.88 2.80 1.96	4	36	0.57 0.67 0.52	1.11 6.03 5.42 1.76
5	20	0.87 0.36 0.39	1.48 1.01 1.50 2.17	5	27	0.47 0.39 0.46	1.95 1.91 3.73 1.93
6	15	0.83 0.83 0.33	1.11 1.25 1.12 2.48	6	24	0.89 0.46 0.43	1.48 1.04 1.55 2.01
7	26	0.42 0.53 0.45	1.84 2.60 1.41 1.96	7	36	0.38 0.65 0.52	1.81 3.86 2.13 1.76
8	26	0.87 0.76 0.45	1.29 2.41 1.86 1.96	8	35	0.58 0.69 0.51	1.18 5.91 5.02 1.77
0907+336							
4	34	1.08 1.26 0.51	1.79 2.22 1.24 1.79	4	35	1.34 1.31 0.51	1.84 2.70 1.46 1.77
6	14	1.38 1.23 0.31	1.40 1.43 2.01 2.58	6	14	1.20 1.01 0.31	4.11 1.12 4.59 2.58
7	13	0.73 0.87 0.29	2.43 2.87 1.18 2.69	7	13	0.39 0.51 0.29	3.01 1.12 3.36 2.69
8	14	1.14 1.24 0.31	1.21 2.78 2.30 2.58	8	14	1.23 1.30 0.31	1.71 1.84 1.08 2.58
9	11	0.76 0.71 0.26	1.05 1.80 1.89 2.98	9	11	1.00 0.44 0.26	4.23 1.04 4.38 2.98
1034+574							
3	47	0.82 0.82 0.57	1.55 2.00 1.29 1.63	3	47	0.79 0.95 0.57	1.16 1.15 1.33 1.63
4	47	0.88 0.63 0.57	2.49 1.48 3.67 1.63	4	47	0.61 0.78 0.57	6.64 1.04 6.92 1.63
6	41	0.73 0.55 0.55	1.15 1.96 1.71 1.69	6	41	0.37 0.49 0.55	1.67 2.19 1.31 1.69
7	25	0.77 0.99 0.44	1.06 1.31 1.39 1.98	7	24	0.81 0.66 0.43	1.47 1.04 1.53 2.01
1212+467							
4	16	1.19 0.90 0.34	1.78 2.23 3.98 2.40	4	16	1.32 0.58 0.34	1.18 1.76 1.49 2.40
5	25	1.16 0.99 0.44	1.12 4.07 3.64 1.98	5	25	0.87 0.84 0.44	1.98 2.83 1.43 1.98
1242+574							
2	43	0.81 0.57 0.56	3.88 3.92 1.01 1.67	2	51	0.64 0.45 0.59	1.12 1.85 1.66 1.60
4	43	1.09 0.95 0.56	2.95 3.69 1.25 1.67	4	51	0.71 0.60 0.59	2.31 2.96 1.28 1.60
5	43	0.54 0.84 0.56	3.41 4.76 1.40 1.67	5	50	0.64 0.73 0.58	2.80 3.16 1.13 1.61
7	42	0.92 1.00 0.55	1.72 1.88 1.10 1.68	7	50	0.86 0.95 0.58	1.16 1.53 1.78 1.61
8	41	0.26 0.39 0.55	1.01 2.70 2.67 1.69	8	47	0.20 0.33 0.57	1.46 6.76 4.63 1.63
1429+249							
3	22	0.29 0.58 0.41	1.17 1.23 1.44 2.08	3	22	0.35 0.59 0.41	1.78 1.30 2.30 2.08
4	14	1.02 0.90 0.31	1.19 2.44 2.91 2.58	4	15	0.84 1.01 0.33	1.76 1.21 2.13 2.48
5	25	0.97 0.65 0.44	1.94 2.33 1.20 1.98	5	25	0.55 0.72 0.44	1.14 1.40 1.60 1.98
8	28	0.85 0.73 0.46	1.37 2.23 1.63 1.90	8	28	0.80 0.81 0.46	1.08 1.84 1.71 1.90
1612+378							
3	31	0.88 0.51 0.49	8.16 1.16 9.47 1.84	3	36	0.56 0.62 0.52	1.22 1.88 2.29 1.76
5	31	0.91 0.79 0.49	1.12 1.44 1.62 1.84	5	36	0.44 0.44 0.52	2.04 4.28 2.10 1.76
8	31	0.73 0.34 0.49	1.17 1.19 1.39 1.84	8	36	0.42 0.50 0.52	1.91 3.18 1.66 1.76

4. CONCLUSIONS

We monitor (for six years) flux changes of AGNs in V, and R bands. The accuracy of differential photometry will be improved by adding more non-variable stars in calculations. For that reason we tested comparison and control stars from the objects vicinity. The tests show that most of the comparison and control stars are useful for differential photometry and only one star with number 8 of source 1242+574 is variable. We compare our data with data from Gaia DR3 catalogue, using equations to transform our V, and R magnitudes to G. We performed least square fit to the data and found good correlations between the data from Gaia DR3 catalogue and our data. The two stars are flagged as variable in Gaia DR3 catalogue. One of them (C_2 - 1722+119) very bright star with brightness of about 13 mag was also flagged as variable in paper Doroshenko et al. (2014). The second one (5 – 1242+574) is located close to the fitting line and tests did not show its variability. The V_{O}, and R_{O} magnitudes of stars (observed) and magnitudes which were the initial values for the differential photometry V_{C}, and R_{C} are in good agreement (see Table 1). We will continue with observations and investigations of Intra Day, Short Term, and Long Term changes in brightness of stars and objects.

Acknowledgements

During the work on this paper authors were financially supported by the Ministry of Education, Science and Technological Development of the Republic of Serbia through the contract No 451-03-68/2022-14/ 200002. We gratefully acknowledge the observing grant support from the Institute of Astronomy and Rozhen National Astronomical Observatory, Bulgarian Academy of Sciences.

References

Abolfathi B., Aguado D. S., Aguilar G. et al.: 2018, *Astrophys. J., Suppl. Ser.*, 235, 42.
Bourda G., Collioud A., Charlot P., Porcas R., Garrington S.: 2011, *Astron. Astrophys.*, 526, A102.
Charlot P., Jacobs C. S., Gordon D. et al.: 2020, *Astron. Astrophys.*, 644, A159.
Chonis T. S., Gaskel M. C.: 2008, *Astron. J.*, 135, 264.
De Diego J. A.: 2010, *Astron. J.*, 139, 1269.
Doroshenko V. T., Efimov Yu. S., Borman G. A., Pulatova N. G.: 2014, *Astrophysics*, 57, 176.
Gaia Collaboration, Vallenari A., Brown A. G. A., et al. 2022, arXiv:2208.00211.
Jovanović M. D.: 2019, *Serb. Astron. J.,* 199, 55.
Jovanović M. D., Damljanović G., Taris F.: 2021, Control stars around quasars suitable for the ICRF – GAIA CRF link, *Publ. Astron. Obs. Belgrade,* No. 100, 253–258.
Razali N. M., Wah Y. B.: 2011, *Journal of Statistical Modeling and Analytics,* 2, 21.
Sartoretti P., Marchal O., Babusiaux C., et al. 2022, *A&A* accepted, arXiv:2206.05725
Tody D.: 1986, *Proceedings SPIE Instrumentation in Astronomy VI,* ed. D.L. Crawford, 627, 733.
Tody D.: 1993, *Astronomical Data Analysis Software and Systems II,* A.S.P. Conference Ser., eds. R.J. Hanisch, R.J.V. Brissenden, & J. Barnes, 52, 173.
van Dokkum P. G.: 2001, *PASP*, 113, 1420.

Proceedings of the XIII Bulgarian-Serbian Astronomical Conference (XIII BSAC) Velingrad, Bulgaria, October 3-7, 2022
Editors: Evgeni Semkov, Milan S. Dimitrijević, Momchil Dechev and Zoran Simić
Publ. Astron. Soc. "Rudjer Bošković" No 25, 2023, 85-95

SEVERAL COSMOLOGICAL NUCLEOSYNTHESIS CONSTRAINTS ON NEUTRINO AND NEW PHYSICS

DANIELA KIRILOVA, MARIANA PANAYOTOVA and EMMANUIL CHIZHOV

Institute of Astronomy and NAO, Bulgarian Academy of Sciences
Tsarigradsko Shosse 72 Blvd., Sofia 1784, Bulgaria
E-mail: dani@astro.bas.bg, mariana@astro.bas.bg

Abstract. We provide short review of the contemporary Big Bang Nucleosynthesis (BBN) and the new more precise measurements of the primordially produced abundances of the light elements D and He-4. We present several cosmological constraints, based on contemporary BBN, on the physical characteristics of neutrino and on beyond Standard Model sterile neutrino.

1. INTRODUCTION

Our motivation to study neutrino is supported by the following main arguments: ***First,*** neutrino is elusive (as it has only weak interactions) but important particle. It is a constituent of the Standard Model (SM) of particle physics. ***Second,*** physical characteristics of neutrino are not yet fully studied. ***Third,*** neutrino has astrophysical importance: sources of neutrino are the Sun and other stars, SN, AGN, early Universe. In particular, neutrino plays important role in early Universe processes. Hence, cosmology constrains neutrino characteristics, its mass, number density, the number of different light neutrino types, oscillation parameters, etc. and, thus, presents complimentary knowledge about neutrino.

Moreover, in recent years combined neutrino oscillations data including reactor experiments +LSND +MiniBooNe +Gallium experiment hint to the possible presence of 1 light right handed neutrino, called further sterile neutrino ν_s, participating in neutrino oscillations with flavor neutrinos with squared mass difference of the order of ~eV^2. Sterile neutrino, as well as neutrino oscillations phenomenon, present physics beyond SM. It is predicted by Grand Unified Theories models. It is used in theoretical models to explain small non-zero neutrino masses. It is also among the preferred particles candidates for Dark

Matter (DM). It has applications in models of large scale structure formation, models of natural baryogenesis through leptogenesis, etc.

In the following sections we will discuss how sterile neutrino is constrained by cosmology, in particular by Big Bang Nucleosynthesis (BBN). It has strong dynamical effect during the early Universe - it increases the expansion rate in case sterile neutrino is brought into equilibrium and, hence, influences light elements production. In case of oscillations with flavor neutrinos it effects nucleons kinetics during pre-BBN epoch and, hence, its mixing parameters are constrained by BBN. We will discuss contemporary cosmological constraints on the number of light neutrino families and the degree of population of the sterile neutrino, We will present recent update to the BBN constraints on neutrino oscillation parameters, and on the temperature of decoupling of light sterile neutrinos.

2. BBN - THE DEEPEST RELIABLE EARLY UNIVERSE PROBE OF NEW PHYSICS

Big Bang Nucleosynthesis is theoretically well established and experimentally and observationally confirmed model explaining the production of the light elements in the early Universe during its hot stage, when the Universe cooled from T~ 1 MeV until 0.1 MeV. Precise data on nuclear processes rates exists from experiments carried out at low energies (10 KeV – MeV). Analyses of nuclear rates for BBN are used like NACRE-I (Angulo et al. 1999) and NACRE-II (Xu et al. 2013). More than 400 nuclear reactions are considered and precise BBN codes are used, namely PArthENoPE, AlterBBN, PRIMAT, etc. (Pisanti et al, 2008, Consiglio et al., 2017, Arbey, 2012, Pitrou et al., 2018, Miele et al. 2011). Observational data on deuterium D, helium He-3, He-4 and lithium Li-7 exists and during the last years the determination of primordially produced D and He has reached high precision.

All 3 parameters, on which primordially produced elements depend have been measured with good precision. Namely, the *baryon-to-photon ratio* η has been measured independently by CMB with high precision: η_{CMB} = (6.104 ± 0.055) 10^{-10}. It is in good agreement with the value obtained in BBN analysis of all the light elements, and in particular its value derived on the basis of D primordial abundance, as D is the best baryometer among the elements produced primordially. Namely, the concordance η range is η_{BBN} = (6.143 ± 0.190) 10^{-10}.at 1 sigma error (Fields et al., 2020). *The number of the effective degrees of freedom of light particles* N_{eff} during the BBN epoch is also known with much higher precision than before $\Delta N_{eff} < 0.3$, where N_{eff} is usually defined by the energy density of light neutrino:

$$\rho_\nu = 7/8(T/T_\nu)^4 N_{eff}\rho_\gamma(T)$$

Thus, ΔN_{eff} may indicate nonstandard interactions, extra relativistic degrees of freedom, exotic physics, i.e. physics beyond SM. *The neutron lifetime* has been also measured with higher accuracy τ = 879.4±0.6 s (see Serebrov et al., 2017, PDG 2022).

There exist remarkable concordance between theoretically predicted and derived from observations abundances of light elements primordially produced. In particular an excellent agreement exists between the predicted for $\eta = 6.10^{-10}$ and derived from observations data abundances of D and He-4. Namely: the theoretically predicted by BBN mass fraction of primordial He-4 is (Pitrou et al., 2018)

$$Y_T = 0.24709 \pm 0.00017,$$

the predicted D to hydrogen ratio is

$$D/H_T = (2.459 \pm 0.036)\ 10^{-5}$$

while their values obtained from observations read correspondingly (Aver et al., 2015, Cooke et al., 2017, Fields et al., 2022 PDG):

$$Y_p = 0.245 \pm 0.003,\ D/H = (2.536 \pm 0.026)\ 10^{-5}$$

Hence, the contemporary BBN is often used as the most deep reliable precision probe for physical conditions in early Universe and as a unique test for new physics.

During the last decade the precision of D and He-4 observational data improved considerably, due to new observations of QAS for D (Cooke et al., 2018, Balashev et al., 2016) and due to the inclusion of He10830 infrared emission line for He-4 that helped for precise primordial abundance determinations.

Primordial He-4 was determined with better than 3% accuracy, mainly due to the inclusion of He 10830 infrared emission line measured in the extremely metal poor galaxy Leo P (Aver et al., 2020):

$$Y_p = 0.2453 \pm 0.0034.$$

This better accuracy of He-4 determinations allows to update and strengthen the Big Bang Nucleosynthesis constraints on physics beyond Standard Model. In the following section we present updates of BBN constraints on several beyond Standard Model neutrino physics based on this uncertainty of Y_p.

3. NEUTRINO COSMOLOGICAL EFFECTS AND BBN CONSTRAINTS ON NEUTRINO CHARACTERISTICS

Here we present updated BBN constraints on beyond SM neutrino physics, corresponding to ^{4}He Y_p/Y_p = 1-3% uncertainty. We consider several beyond SM physics models and present updated constraints on new physics. Namely: first we discuss the recent cosmological constraints on the number of the effective degrees of freedom of light particles during the BBN epoch; Then we discuss the model of BBN with neutrino oscillations and present the updated stringent BBN constraints on neutrino oscillations parameters recently derived (Kirilova&Panayotova, 2022). Next we discuss BBN constraints on lepton asymmetry in BBN model with neutrino oscillations and lepton asymmetry (Kirilova, 2012, 2013, 2019). Finally we present updated BBN based constraints on the freezing temperature of light sterile neutrinos (Kirilova, E.Chizhov, 2019, Kirilova, E. Chizhov, V.Chizov, 2020).

3.1. Cosmological constraints on the number of light neutrino types

Additional light particles into equilibrium increase universe expansion rate thus influencing BBN, hence BBN constrains the effective number of relativistic species N_{eff}.

BBN and recently determined primordial He-4 and D (Pitrou et al., 2018, Cooke et al., 2017) lead to the following constraint:

$$N_{eff}= 2.88\pm0.154 \ (95\%).$$

A maximum likelihood analysis on η and N_{eff} based on He-4 and D abundances finds (Fields et al. 2020):

$$N_{eff}= 2.843\pm0.27 \ (95\%) \text{ and } \eta=6.09\pm0.055 \ (95\%).$$

Recent cosmological constraint on N_{eff}, based on BBN+CMB (Yeh et al., 2022) is even more restrictive:

$$N_{eff} = 2.898 \pm 0.141 \qquad N_{eff} < 3.18 \text{ at } 95\% \text{ CL}$$

Combined constraints based on updated Planck CMB data (Pitrou, 2018) give:

$$N_{eff} = 3.01\pm0.15 \text{ at } 95\% \text{ CL}.$$

BBN and CMB neutrino numbers are *consistent* with the standard cosmological model value $N_{eff} = 3.045$ within uncertainties. Still, non-zero $\Delta N_{eff} = N_{eff} - 3.045$ may indicate extra relativistic component, like light right handed neutrino usually called sterile neutrino.

In subsection 3.4 we will use $\delta N_{eff} < 0.2$ for obtaining BBN constraints on sterile neutrino decoupling and on new beyond SM interactions. This constraint corresponds to 1% He-4 uncertainty discussed in the following subsection 3.2, where we obtain BBN limits on the electron-sterile neutrino oscillations.

3.2. BBN constraints on neutrino oscillations

In BBN with $\nu_e \leftrightarrow \nu_s$ proceeding ***before active neutrino decoupling*** oscillations can exite into equilibrium the sterile neutrino state (Dolgov, 1988, Barbieri, Dolgov, 1990,1991) and lead to higher expansion rate. While in case neutrino oscillations proceed ***after the electron neutrino decoupling*** neutrino energy distribution and the density of electron neutrino may considerably differ from the equilibrium ones in standard BBN. This leads to different nucleon kinetics and modifies BBN element production (Kirilova, 1988, Kirilova, Chizhov, 1996, 1997). Nucleons evolution in the pre-BBN period in the presence of late $\nu_e \leftrightarrow \nu_s$ effective after neutrino freezing, was numerically analyzed for different sets of oscillation parameters, accounting for the distortion in electron neutrino energy distribution due to neutrino oscillations in a cycle of papers. The primordially produced He-4 was calculated and compared with its observational uncertainty to provide BBN constraints on neutrino oscillation parameters – squared mass differences δm^2 and mixing $\sin^2 2\theta$.

In the 90ies ^{4}He was known with ~3-7% accuracy and we have obtained BBN constraints on δm^2 and $\sin^2 2\theta$ corresponding to 3-7% overproduction of ^{4}He (Kirilova, Chizhov, 1998, 2000). These constraints were more precise, namely orders of magnitude more stringent than previous ones, due to the precise account for energy spectrum distortion caused by late neutrino oscillations. They excluded LOW active-sterile solution to the solar neutrino problem in addition to the already excluded LMA active-sterile solution to the solar neutrino problem in previous publications (Kainulainen, 1990, Barbieri, Dolgov, 1991, Enquist et al., 1992)

Recently we have provided numerical analysis of more than 100 BBN models with electron-sterile neutrino oscillations with different sets of oscillation parameters (Kirilova, Panayotova, 2022). We have used updated data on baryon density and the neutron life time. We have followed numerically the evolution of the neutron-to-proton ratio in these beyond SM models from the epoch of neutrino freezing until the formation of elements. We have calculated the production of He-4 and obtained different isohelium contours, on the basis of which we have obtained updated BBN constraints on neutrino $\nu_e \leftrightarrow \nu_s$ oscillations parameters, based on 1% ^{4}He uncertainty.

In Fig.1 combined iso-helium contours for 1, 3 and 5% ^{4}He overproduction, accounting for all oscillations effects on BBN, for initial population $\delta N_s = 0$, for non-resonant $\delta m^2 < 0$ and resonant $\delta m^2 > 0$ cases, are given. The lowest contour presents the updated more stringent BBN constraints on $\nu_e \leftrightarrow \nu_s$ oscillations parameters.

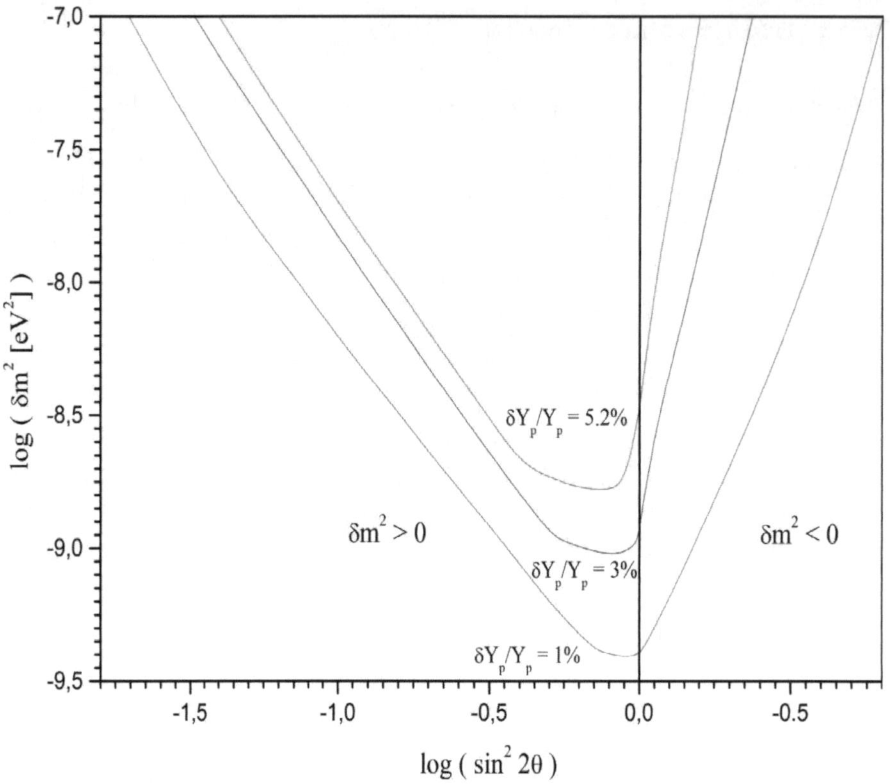

Figure 1: BBN constraints on $\nu_e \leftrightarrow \nu_s$ neutrino oscillation parameters, corresponding to different He-4 uncertainty.

3.3. BBN with lepton asymmetry and neutrino oscillations

Lepton asymmetry L is usually defined as the difference between the number of leptons and antileptons over the number of photons:

$$L = (n_l - n_{\bar{l}}) / n_\gamma$$

In case of equilibrium it can be also expressed through the chemical potentials ξ, as

$$L = \sum_i \frac{1}{12\zeta(3)} \frac{T_{\nu_i}^{\;3}}{T_\gamma^{\;3}} (\xi_{\nu_i}^3 + \pi^2 \xi_{\nu_i})$$

Degenerate BBN has been studied in numerous papers since first analysis of (Wagoner et al., 1967). It is known that L has **dynamical effect**, it increases the radiation energy density, leading to faster expansion, delaying matter/radiation equality epoch and, thus, influencing BBN, CMB, and the evolution of Large Scale Structure. This effect can be described in terms of change of N_{eff}:

$$\Delta N_{eff} = 15/7((\xi/\pi)^4 + 2(\xi/\pi)^2)$$

L present in the electron neutrino sector $|L_{\nu e}| > 0.01$ exerts a ***direct kinetic effect*** – it influences the neutron-proton kinetics during pre-BBN epoch, influencing BBN.

This effect is L sign dependent. Due to neutrino flavor oscillations degeneracies in different neutrino sectors equalize (Dolgov et al., 2002, Mangano et al., 2011). BBN conservative constraint for all neutrino sectors reads is

$$|\xi| < 0.1, \text{ i.e. } |L| < 0.07.$$

In case of BBN with late electron-sterile neutrino oscillations there exists ***indirect kinetic effect*** of L: L in the range $10^{-8} < L << 0.01$, that is too small to effect directly BBN kinetics, influences BBN via oscillations. It effects neutrino evolution, its number density, spectrum distribution, oscillations pattern and hence n/p kinetics and BBN (Kirilova, Chizhov, 1998, Kirilova, 2012, 2019). Depending on its value it can suppress neutrino oscillations (Foot&Volkas, 1995, Kirilova, Chizhov, 1998) or lead to their resonant enhancement (Kirilova, Chizhov, 1998, Kirilova, 2012). We have numerically studied that interplay and determined the parameter range for which L is able to enhance, suppress or inhibit oscillations (Kirilova, 2012, 2019). Recently we have updated our results (Kirilova, 2019). We have found that $L \sim 10^{-7}$ enhances oscillations, strengthens the BBN bounds on them while larger L suppresses oscillations. Thus L, depending on its value, can relax or strengthen BBN constraints on neutrino oscillations. Full suppression of oscillations with squared mass difference δm^2 is possible for

$$L > (0.01\delta m^2)^{3/5}$$

These results can be applied to solve the so called dark radiation problem. Namely, several small base line experiments: reactor experiments, LSND, MiniBooNe, Gallium expt, SAGE and recently Ice Cube hint to a presence of sterile neutrino participating into oscillations with flavor neutrinos with large mixing and mixing squared mass difference in the range $\Delta m^2_{41} = 10^{-2}$ to several eV^2 (Kopp et al., 2011, Dentler et al., 2018, Aartsen et al., 2020). However, oscillations with ν_s with eV^2 mass and large mixing lead to thermalization of ν_s at BBN epoch, i.e. to one full additional light neutrino species into equilibrium. Fully hermalized light inert state is not allowed by BBN ($\delta N_{eff} \sim 0.2$).

Different solutions to the DR problem have been discussed. We have proposed the following solution (Kirilova, 2012, 2013): Large enough L present during BBN, capable to suppress neutrino oscillations, will prevent the thermalization of ν_s, and thus avoid BBN cosmological constraints. See also the analyses by (Mirizzi et al., 2012, Hannestad et al., 2012).

Therefore, from our analysis (Kirilova, 2019) we can conclude that: In case sterile neutrino of mass difference 0.1 eV^2 exists, (as indicated by some reactor experiments) to prevent neutrino oscillations with the active neutrinos it is required that $|L| > 0.016$. In case of mass difference of 1 eV^2 bigger L is needed $|L| > 0.063$, for mass difference of 1.3 eV^2 correspondingly, L should be $|L| > 0.074$. Thus, the detection of sterile neutrino oscillations and their squared mass difference might be used to put lower bound on L present during the BBN epoch.

Recent EMPRESS survey of extremely metal poor systems *(Matsumoto et al. 2022)* obtained 3σ lower primordial He-4 value than the predicted:

$$Y_p = 0.2379^{+0.0031}_{-0.003}$$

This result can be interpreted as a hint for L in the electron neutrino sector:

$$\xi_{\nu e} = 0.039 \pm 0.014 \text{ at } 3\sigma.$$

On the basis of our analysis it is possible to estimate that L corresponding to this $\xi_{\nu e}$, L~0.027±0.01, is capable to suppress neutrino oscillations with

$$\delta m^2 \leq 0.3 eV^2$$

For bigger mass differences, predicted by some neutrino oscillations experiments, L~ 0.03 is able only to suppress partially oscillations. To estimate the exact degree of thermalization of the sterile state in this case more sophisticated numerical analysis is necessary.

It is intriguing that the Hubble tension can be resolved for similar value of ξ~0.04 and $0.3<\delta N_{eff}<0.6$ (Selo, Toda, 2021).

Hence, the preference of the solution of dark radiation puzzle by non-zero L increases.

3.4. BBN constraint on sterile neutrino decoupling

Using BBN constraint, (Pitrou et al. 2018) $\delta N_{eff} < 0.3$ at BBN epoch and recent constraint $\delta N_{eff} < 0.2$ and entropy conservation, we have calculated lower limits of the decoupling temperature T_d of sterile neutrinos (Kirilova, E. Chizhov, 2019). See also (Kirilova, E. Chizhov, V. Chizhov, 2020). We have updated the analysis using the study of the effective degrees of freedom in the early universe provided in ref. (Husdal, 2016).

For the case of 3 light sterile neutrino types and $\delta N_{eff} < 0.3$ we have determined $T_d > 200$ MeV. In case of 2 light sterile neutrino $T_d > 180$ MeV, for 1 light sterile neutrino $T_d > 140$ MeV.

Using the more stringent BBN constraint, $\delta N_{eff} < 0.2$, analogously we have obtained for 3, 2 and 1 light sterile neutrino, correspondingly: $T_d > 1600$ MeV, $T_d > 200$ MeV, $T_d > 170$ MeV. These are slightly lower values than previously estimated.

Comparing this decoupling temperature with the electron neutrino decoupling temperature 2 MeV, one can get an insight of the coupling constant of the eventual interactions of the sterile neutrinos. For example constraints can be obtained on sterile neutrino coupling to Z' gauge bosons in models of superstrings, extended technicolor, etc.

For example, in case sterile neutrinos interact with the chiral tensor particles (CTP), proposed in a model of beyond SM physics (M. Chizhov, 1993), they may be produced through chiral tensor particles exchange during BBN. Then, using the calculated T_d, BBN constraints on the new chiral tensor interactions can be obtained. Our analysis (Kirilova, Chizhov, 2019, Kirilova, Chizhov, Chizhov, 2020) shows that CTP interactions either are milliweak or weaker, or CTP do not interact with sterile neutrinos.

4. CONCLUSIONS

We use recent precise determinations of the primordial abundances of D and He-4 with 1% precision and the stringent cosmological constraints on the number of the effective degrees of freedom of light particles during the BBN epoch $\Delta N_{eff} \leq 0.2$ to obtain BBN constraints on several characteristics of neutrino.

We analyze the model of BBN with $\nu_e \leftrightarrow \nu_s$ neutrino oscillations and derive stringent updated BBN constraints on neutrino oscillations parameters corresponding to the present accuracy of determination of the primordially produced He-4.

The role of lepton asymmetry L was studied. Large L may provide relaxation of BBN constraints on oscillations, by suppressing oscillations. New stringent cosmological constraints on L were derived in the model of degenerate BBN with neutrino oscillations.

A solution to the dark radiation problem is discussed: in model of BBN with neutrino oscillations and large enough L, L is capable to suppress oscillations, and thus to hinder thermalization of the sterile neutrino due to oscillations and avoid BBN constraints on neutrino oscillations with big mass differences and mixings. The values of L necessary for solving the dark radiation problem for different mass differences are estimated.

Cosmological constraints on the decoupling T of sterile neutrino were obtained

In case of any new interactions with right-handed neutrinos stringent BBN constraints may be obtained on their interaction strength.

Neutrino cosmological influence may be considerable. Hence, studying its cosmological effects provides important complimentary and unique physical knowledge. This is precious knowledge having in mind that due to its extremely week interactions sterile neutrino is difficult to be studied experimentally.

Acknowledgements

We thank the organizers of the XIII Bulgarian-Serbian conference for the pleasant and inspiring atmosphere of the meeting. We thank M. Chizhov for the valuable help during the determination of the decoupling temperatures.

References

Aartsen M. et al.: 2020, *Phys. Rev. Lett.*, 125(14),141801.
André de Gouvêa et al.: 2022, arXiv 2204.09130.
Angulo C. et al.: 1999, *Nucl. Phys. A*, 656, 3.
Arbey A.: 2012, *Comput. Phys. Commun.*, 183, 1822.
Aver E., Olive K., Skillman E.: 2015, *JCAP*, 7, 011.
Aver E. et. al.: 2021, *JCAP,* 3, 027 .
Balashev et al.: 2016, *MNRAS*, 458(2), 2188.
Barbieri R., Dolgov A.: 1990, *Phys. Lett. B,* 237, 440.
Barbieri R., Dolgov A.:1991, *Nucl. Phys. B*, 349, 743.
Sz. Borsanyi et al.: 2017, *Nature*, arXiv:1606.07494v2
M. Chizhov: 1993, *Mod. Phys. Lett. A*, 8, 2753.
Consiglio et al.: 2017, arXiv 1712.04378.
Cooke R., Pettini M., Steidel C.: 2017, *Astrophysical Jour*nal, 855(2), 102.
Dentler et al.: 2018, Updated Global Analysis of Neutrino Oscillations with eV-Scale Sterile Neutrinos, *JHEP*, 08, 010.
Dolgov A.: 1981, *Sov. J. Nucl. Phys.*, 33, 700.
Dolgov A.: 2002, *Phys. Rept.*, 370, 333.
Dolgov A. et al., 2002, *Nucl. Phys B*, 632, 363.
Enqvist K., Kainulainen K., Thompson M.: 1992, *Nucl. Phys.B,* 373, 498.
Fields et al.: 2022, *Big Bang Nucleosynthesis*, Particle Data Group
Fields et al.: 2020, *JCAP*, 2020, 3, 010
Fields et al.: 2020, *JCAP*, 2020, 11, E02.
Foot R., R. Volkas: 1995, *Phys.Rev.Lett.*, 75, 4350.
S. Hannestad, I. Tamborra, T.Tram: 2012, *JCAP*, 1207, 025.
Hsyu T. et.al., arXiv:2005.12290 accepted *ApJ* 2020
Husdal L.: 2016, *Galaxies*, 4, 78
Kainulainen K.: 1990, *Phys.Lett.B*, 244, 191-195
Kirilova D., Panayotova M.: 2022, to be published.
Kirilova D., E. Chizhov, V. Chizhov: 2020, *Journal of Physics: Conference Series*, 1668(1), 012022
Kirilova D.: 2019, PoS CORFU2018, v. 48
Kirilova D.: 2013, *Hyperfine Interact.*, 215,111.
Kirilova D.: 2012, *JCAP*, 06, 007.
Kirilova D.: 2007, *IJMPD*, 16(7), 1
Kirilova D.: 2004, *Int. J. Mod. Phys. D,* 13, 831.
Kirilova D.: 1988, *JINR,* preprint E2-88-301.
Kirilova D., E. Chizhov, 2019, *IJ Mod. Phys. A*, 34, N11, 1950065
Kirilova D., V. Chizhov: 2017, *IJ Modern Physics Letters A*, 32, 1750187.
Kirilova, D., Chizhov, M.: 1996, *Neutrino*, 96, 478.
Kirilova, D., Chizhov, M.:1997, *Phys. Lett. B*, 393, 375.
Kirilova D., M. Chizhov: 1998a, *Phys.Rev.D*, 58, 073004,
Kirilova D., M. Chizhov: 1998b, *Nucl. Phys. B*, 534, 447.
Kirilova D., M. Chizhov: 2000a, *MNRAS*, 314, p.256.
Kirilova D., M. Chizhov: 2000b, *Nucl.Phys.B*, 591,457.
Kopp J., Maltoni M., Schwetz T.: 2011, *Phys.Rev.Lett.*, 107, 091801.
Mangano G.,Serpico P.: 2011, *PLB*, 701, 296.
Mangano G., G. Miele, S. Pastor, O.Pisanti, S. Sarikas: 2011, *JCAP*, 11, 035.

Matsumoto A. et al., EMPRESS VIII: 2022, arXiv 2203.09617
A. Mirizzi et al.: 2012, *PRD*, 86, 053009.
Pisanti O. et al.: 2008, *Comput. Phys. Commun.*, 178, 956.
Pitrou C., Coc A., Uzan J.-P., Vangioni E.: 2018, *Physics Reports*, 04, 005.
Serebrov A. et al.: 2017, arXiv 1712.05663, *Phys.Rev.C*, 97, 055503.
Valerdi M. et.al.: 2019 *Astrophys. J.*, 876(2), 98.
Wagoner R., Fowler W., Hoyle F.: 1967, *Astrophys. J.Supp.*, 148, 3.
Xu Y. et al.: 2013, *Nucl. Phys. A*, 918, 61.
Yeh Tsung-Han, J. Shelton, K. Olive, B. Fields: 2022, *JCAP*, 10, 046.

Proceedings of the XIII Bulgarian-Serbian Astronomical Conference (XIII BSAC) Velingrad, Bulgaria, October 3-7, 2022
Editors: Evgeni Semkov, Milan S. Dimitrijević, Momchil Dechev and Zoran Simić
Publ. Astron. Soc. "Rudjer Bošković" No 25, 2023, 97-107

WAVELET COHERENCE OF TOTAL SOLAR IRRADIANCE AND ATLANTIC CLIMATE

VASIL KOLEV[1] and YAVOR CHAPANOV[2]

[1]*Institute of Information and Communication Technologies, Bulgarian Academy of Sciences Acad. Georgi Bonchev Str., Block 2, Sofia 1113, Bulgaria*
[2]*Climate, Atmosphere and Water Research Institute, Bulgarian Academy of Sciences*
Acad. Georgi Bonchev Str., Block 6, Sofia 1113, Bulgaria
E-mail: kolev_acad@abv.bg, vasil.kolev@iict.bas.bg, yavor.chapanov@gmail.com

DEDICATED TO THE MEMORY OF PROF. MILCHO TSVETKOV

Abstract. The oscillations of climatic parameters of North Atlantic Ocean play important role in various events in North America and Europe. Several climatic indices are associated with these oscillations. The long term Atlantic temperature anomalies are described by the Atlantic Multidecadal Oscillation (AMO). The Atlantic Multidecadal Oscillation also known as Atlantic Multidecadal Variability (AMV), is the variability of the sea surface temperature (SST) of the North Atlantic Ocean at the timescale of several decades. The AMO is correlated to air temperatures and rainfall over much of the Northern Hemisphere, in particular in the summer climate in North America and Europe. The long-term variations of surface temperature are driven mainly by the cycles of solar activity, represented by the variations of the Total Solar Irradiance (TSI). The frequency and amplitude dependences between the TSI and AMO are analyzed by wavelet coherence of millennial time series since 800 AD till now. The results of wavelet coherence are compared with the detected common solar and climate cycles in narrow frequency bands by the method of Partial Fourier Approximation. The long-term coherence between TSI and AMO can help to understand better the recent climate change and can improve the long term forecast.

1. INTRODUCTION

Many studies have shown coherence between TSI and other important time series parameters (Benevolenskaya et al., 2013). In (Chapanov, 2021) the narrow frequency bands of the cycles of Atlantic multidecadal oscillation (AMO) are compared with corresponding oscillations of total Solar Irradiance (TSI). By using the method of Partial Fourier Approximations (PFA) between interannual and decadal cycles, the correlation between the AMO and the TSI are shown for periodicities between 48 and 233 years. The correlation between TSI and AMO is also considered in (Velasco et al., 2008; Swingedouw et al., 2010; Shindell et al., 2001).

Wavelet transforms are one well-known approach for time series processing. Wavelets are basis functions possess the important properties – orthogonality, compact support, optimal time and frequency resolution, and so on. When used in time series analysis wavelet transforms capture the non-stationarity in the spectrum very well. The analysis of different type signals has the ability simultaneously to extract local spectral and temporal information by decomposition of different resolutions. Therefore, they provide a space-scale or time-frequency analysis, they are useful a tool for the investigation of time series.

The continuous wavelet transform (CWT) allows for the analysis of non-stationary signals at multiple scales and can analyze a signal with good resolution in both time and frequency by using continuous wavelets (for example - Morlet, Paul, DoG, Mexican Hat). The obtained wavelet coefficients are functions of the scale and positions as they are computed as sums over all time of the signal multiplied by scaled and shifted versions of the wavelet. In time series analysis using the CWT the continuous Morlet wavelet is preferred because it has optimal joint time-frequency concentration. The mathematics of the CWT was first introduced in (Goupillaud et al., 1984). In the time-frequency analysis it has advantages because it is Gaussian-shaped in the frequency domain and can retain the temporal resolution of the signal. Its convolution is implemented by using the FFT which is computationally efficient. By using continuous wavelet (Xu et al., 2021) is construct the TSI reconstruction model and the monthly total solar irradiance (TSI) time series for 1907 to 1978 years. The results of CWT for the period from 1979 to 2015 years show that the TSI and sunspot number all have significant and stable oscillation periods of 9□13 years and intermittent oscillation periods of 2□6 months only during the times of intense solar activity.

In this paper, we focus on the TSI reconstruction based on sunspot number and AMO index. First, we consider in sections 1.1, 1.2, and 2 the two methods for processing time series – the partial Fourier approximation and the estimation of the wavelet spectrum, and their coherence. Second, in sections 3 and 4 the TSI and AMO are introduced in brief. Finally, in section 5 we study how to examine the solar periodicities that could possibly be related with the AMO for the time period 800 – 2010 (Wang et al., 2017) and the coherence with the TSI for the time period 850 – 2010 (Fröhlich et al, 2018).

1.1 Partial Fourier Approximation

The periodic variations of time series are determined by the Method of Partial Fourier Approximation (PFA). The time series of oscillations from a given frequency band are calculated as a superposition of two neighbor Fourier harmonics, whose coefficients are estimated by the Least Squares (LS) Method. The details of this method are described in (Chapanov et al., 2015). Shortly, the partial Fourier approximation $F(t)$ of discrete data is given by:

$$F(t) = f_0 + f_1(t - t_0) + \sum_{k=1}^{n} a_k \sin\frac{2\pi k}{P_0}(t - t_0) + \sum_{k=1}^{n} b_k \cos\frac{2\pi k}{P_0}(t - t_0) \quad (1)$$

where P_0 is the period of the first harmonic, t_0 - the mean epoch of observations, f_0, f_1, a_k and b_k are unknown coefficients and n is the number of harmonics of the partial sum, which covers all oscillations with periods between P_0/n and P_0. The application of the LS estimation of Fourier coefficients needs at least $2n + 2$ observations, so the number of harmonics n is chosen significantly smaller than the number N of sampled data f_i. The small number of harmonics n yields to LS estimation of the coefficient errors. This estimation is important difference with the classical Fourier approximation. The other difference is the possibility of arbitrary choice of the period of first harmonic P_0, instead of the observational time span, so the estimated frequencies may cover the desired set of real oscillations. This method allows a flexible and easy separation of harmonic oscillations into different frequency bands by the formula:

$$B(t) = \sum_{k=m_1}^{m_2} a_k \sin k\frac{2\pi}{P_0}(t - t_0) + \sum_{k=1}^{n} b_k \cos k\frac{2\pi}{P_0}(t - t_0) \quad (2)$$

where the desired frequencies ω_k are limited by the bandwidth

$$\frac{2\pi m_1}{P_0} \le \omega_k \le \frac{2\pi m_2}{P_0}, \quad (3)$$

After estimating the Fourier coefficients, it is possible to identify a narrow frequency band presenting significant amplitude, and defining a given cycle. Then this cycle can be reconstructed in time domain as the partial sum limited to the corresponding frequency bandwidth.

1.2 Continuos Wavelet Transform

The Morlet wavelets are a one – parameter family of functions and consist of a complex exponential modulated by a Gaussian (Cohen, 2019),

$$\psi(t) = e^{i2\pi f_0 t} e^{-\frac{4\ln(2)t^2}{h^2}} \quad (4)$$

where h - is full-width at half-maximum (FWHM), and in seconds is the distance in time between 50% gain before the peak to 50% gain after the peak. Therefore, for a given square – integrable signal $s(t) \in L^2(R)$ in the time domain $(-\infty, \infty)$, analyzing with the continuous wavelet transform (CWT) is the inner product of the signal with a series of daughter wavelet functions and is defined with an equation, i.e.

$$W(a,b) = \langle s, \psi \rangle = \frac{1}{\sqrt{a}} \int_{-\infty}^{\infty} s(t) \psi \left(\frac{t-b}{a}\right) dt, \tag{5}$$

where $L^2(R)$ denotes the Hilbert space of measurable, square – integrable one-dimensional signals, $\psi(t)$ represents a mother wavelet, a is a dilation/compression scale factor that determines the characteristic frequency, b - the translation. Therefore, by reducing the scaling parameter a we can reduce the support of the wavelet in space, which covers higher frequencies, and vice versa. This means that

$1/a$ is a measure of frequency. The parameter b indicates the location of the wavelet window along the space axis.

Wavelet dilation increases the CWT's sensitivity to long time-scale events, and wavelet contraction increases its sensitivity to short time-scale events. This changing of the pair (a, b) enables computation of the wavelet coefficients $W(a, b)$ on the entire space-frequency plane. Reconstruction of a signal from the wavelet coefficients is obtained by the expression:

$$s(t) = \frac{1}{C_\psi} \int_{-\infty}^{0} \int_{0}^{\infty} W(a,b) |a|^{-\frac{1}{2}} \psi \left(\frac{t-b}{a}\right) \frac{dadb}{a^2} \tag{6}$$

where the Fourier transform of the mother wavelet $\hat{\psi}(\omega)$ needs to satisfy the following admissibility condition,

$$C_\psi = \int_{-\infty}^{\infty} \frac{|\hat{\psi}(\omega)|^2}{|\omega|} d\omega < +\infty,$$

where C_ψ is the *admissibility constant* depending on the chosen wavelet. It

follows that $\hat{\psi}(\omega)$ is a continuous function, so that the finiteness of C_ψ implies that $\hat{\psi}(0) = 0$, has no zero frequency component and the mean value of $\psi(t)$ in the time domain is zero, i.e. $\int_{-\infty}^{\infty} \psi(t) dt = 0$. The wavelet $\psi(t)$ is a window function that simultaneously enables the possibility of time–frequency localization and has finite energy

$$E = \int_{-\infty}^{\infty} |\psi(t)| dt < +\infty.$$

More details for the CWT are given in (Sadowsky, 1994, 1996; Najmi et al., 1997). The absolute values of the CWT show a function of time and frequency that is *a scalogram.* Since the "*effective support*" of the wavelet at scale a is proportional to a, these edge – effects also increase with a. The region in which the transform suffers from these edge effects is called cone of influence (COI). COI is the set of all t included in the effective support $supp_{eff}(b - aK, b + aK)$ of the wavelet at a given position and scale. At each scale, the COI determines the set of wavelet coefficients influenced by the value of the signal at a specified position. Usually, the COI includes the line and the shaded region from the edge of the line to the frequency (or period) and time axes as it shows areas in the scalogram potentially affected by *edge-effect artifacts* (Mallat, 1999). It is important to note that these effects arise from areas where the stretched wavelets extend beyond the edges of the observation interval. Therefore, the information provided by the scalogram is an accurate time-frequency representation of the data within the unshaded region delineated by the white line, while outside the white line in the shaded region, information in the scalogram can be treated as suspect.

2. WAVELET COHERENCE AND PHASE

The wavelet coherence between time series is related to the periodic phenomena in time series. It is similar to a bivariate correlation coefficient and measures the degree and extent of co-movement between two-time series variables but in the time-frequency domain. A wavelet coherency could identify both frequency bands and time intervals where the time series relate. It is defined using smoothing in both time and scale, with the amount of smoothing dependent on both the choice of wavelet and the scale. Given two time series X and Y, with wavelet transforms $W_n^X(s)$ and $W_n^Y(s)$ the cross-wavelet spectrum is defined as

$$W_n^{XY}(s) = W_n^X(s)\, W_n^{Y*}(s) \tag{7}$$

where n is the time index, s is the scale, and (*) indicates complex conjugate. The *wavelet-squared coherency* is defined as the absolute value squared of the smoothed cross-wavelet spectrum, normalized by the individual wavelet power spectra (Velasco et al., 2008),

$$R_n^2(s) = \frac{S(s^{-1}|W_n^{XY}(s)||^2)}{S(s^{-1}|W_n^X(s)|^2)S(s^{-1}|W_n^X(s)|^2}, \quad 0 \le R_n^2(s) \le 1 \tag{8}$$

where $\langle\cdot\rangle$ indicates smoothing in both time and scale, S is a smoothing operator. The factor s^{-1} is used to convert to an energy density. Finally, it is noted that because the wavelet transform conserves variance, the wavelet coherency is an accurate representation of the (normalized) covariance between the two time series.

The wavelet-coherency phase difference given by

$$\phi_n(s) = \tan^{-1}\left(\frac{\operatorname{Re}(s^{-1}W_n^{XY}(s))}{\operatorname{Im}(s^{-1}W_n^{XY}(s))}\right) \tag{9}$$

where the smoothed real and imaginary parts is already been calculated in (8). Both $R_n^2(s)$ and $\phi_n(s)$ are functions of the time index n and the scale s. The smoothing in (8) and (9) was been done using a weighted running average (or convolution) in both the time and scale directions.

It should be note that wavelet coherency depends on an arbitrary smoothing function. Nevertheless, a ''natural'' width of the smoothing function in both time and Fourier space can be provides use the Morlet wavelet function. The time smoothing uses a filter given by the absolute value of the wavelet function at each scale, normalized to have a total weight of unity. The significance level of the wavelet coherence is also determined using the Monte Carlo simulations.

The advantage of wavelet coherence over wavelet power is that it shows statistical significance only in areas where the series involved actually share significant periods. Its disadvantage may be that it needs careful fine-tuning in computation and plotting.

3. TOTAL SOLAR IRRADIANCE

The TSI (Fligge et al., 2000, Foukal, 2012) is the value of the integrated solar energy flux over the entire spectrum arriving at the top of the terrestrial atmosphere at the mean Sun - Earth distance (the astronomical unit, AU). Mathematical calculations, as well as satellite observations indicate an average value of 1367 ± 4 W/m^2. After distribution of the solar irradiance over the planetary atmosphere, the average solar radiation at the top of the atmosphere is 1/4 of this: 342 W/m^2. The Earth's planetary albedo of 0.3 further reduces the incoming radiation to 239 W/m^2. Upon entering the atmosphere solar irradiance wavelengths shorter than 300 nm are absorbed in the stratosphere and above. Satellite measurements of the TSI started with NIMBUS-7 launched in November 1978 and has been carried out by their successors.

The TSI has been widely used as a characteristic of solar variability as its variations can be separate into a cyclic and a long-term component that is calculated using sunspots and faculae as the main contributors.

4. ATLANTIC MULTIDECADAL OSCILATION (AMO) INDEX

The AMO (Enfield et al., 2001) is ocean oscillations system which sets the character of regional changes in oceanic environment and is the theorized variability of the sea surface temperature (SST) or a climate mode in the North Atlantic Ocean (NAO) from 0° to 70° N, occurring at the time scale of several decades. Its index is a measure of basinwide sea surface temperature variation in the North Atlantic, adjusted to remove trends in anthropogenically forced warming.

Like other modes of variability (e.g., El Niño Southern Oscillation), the AMO has impacts on a large geographic scale via atmospheric teleconnections, and has been hypothesized to have an influence on a range of North Atlantic fisheries and ecosystems. AMO has been identified as a coherent mode of natural variability occurring in the North Atlantic Ocean with an estimated period of 60-80 years. It is based upon the average anomalies of sea surface temperatures (SST) in the North Atlantic basin, typically over 0-80N. To remove the climate change signal from the AMO index, users typically detrend the SST data at each grid point or detrend the spatially averaged timeseries.

5. RESULTS

5.1 Solar influence on centennial and decadal TSI and AMO variations

All solar influences of TSI and AMO variations estimated by using of partial Fourier approximation are obtained in (Chapanov, 2021). The correlation periodicities between AMO and TSI for 850 – 2011 are presented in Table 1. The agreement between time series with these oscillations is excellent with a high correlation between them for the time interval 1100-2010 yrs.

Table 1: The correlation periodicities for 850 – 2011 years between AMO and TSI

	Correlations	
	Decadal, years	Centennial, years
1	72.4-77.3	193.2-232.8
2	64.4-68.2	144.9-165.6
3	58.0-61.0	115.9-128.8
4	48.3-54.4	

5.2 CWT of TSI and AMO

The used time series of TSI and AMO long-term variations cover 1161years and 1211 years overlapped time interval for the period 850.0 – 2011 yrs and 800.0 – 2011 yrs, respectively shown of fig.1. The power spectra of the individual AMO and TSI time series with CWT using the Morlet wavelet is shown in fig. 2. The vertical and horizontal axes show the scale (years) bands and a period in years, respectively. The significant co-movements are marked by dark red colors on the top of the sidebar. In contrast, colder blue colors signify lower dependence between the series. Cold regions beyond the significant areas represent time and frequencies with no dependence in the series. The thin black line and the gray line cone delineate the cone of influence (COI) region where the estimates of wavelet coefficients are statistically insignificant at a 5% level of significance. This is estimate by Monte Carlo simulation with phase-randomized surrogate series. Lightly shaded zones inside the COI are ruled out from the analysis, due to the potential presence of edge artifacts.

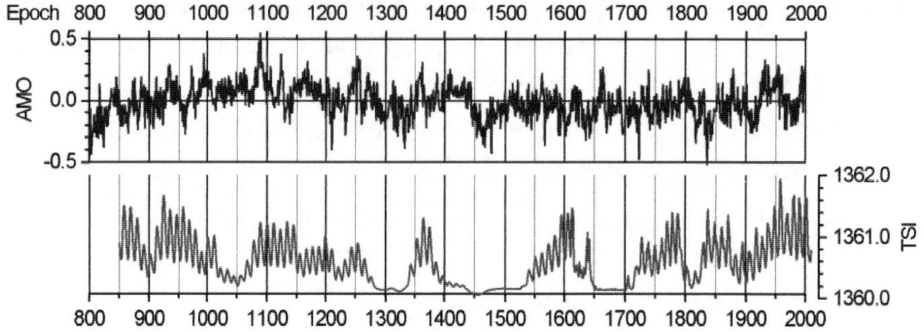

Figure 1: AMO and TSI data.

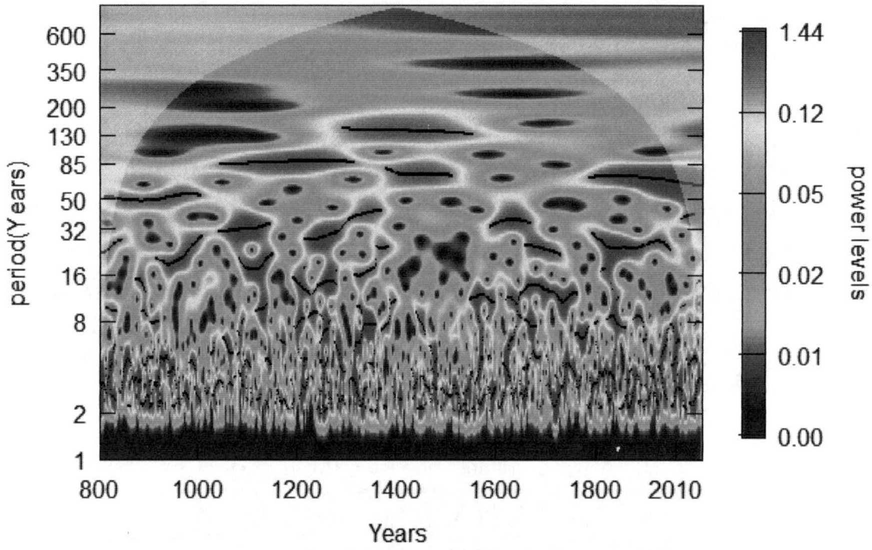

(a)

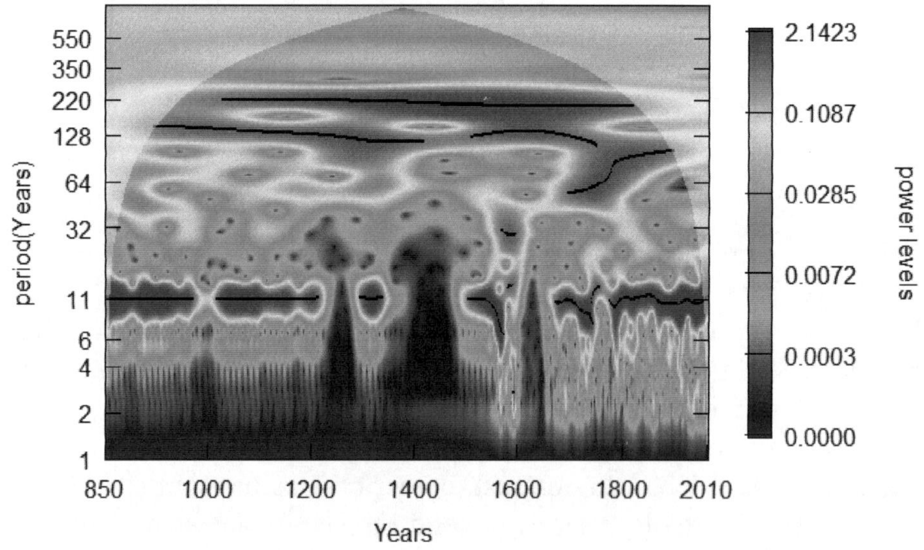

(b)

Figure 2: Wavelet spectrums of (a) AMO and (b) TSI.

Outside the COI, statistically significant regions at the 0.05 level are enclosed by a thick black contour. On the *y*-axis, each series is decomposed into 8 – time scales, where the shortest one (i.e., 1/2 – 1 years) denotes the highest frequency band and the longest one (i.e., 256 – 512 years) denotes the lowest frequency band.

From the obtained CWT results shown in fig. 2 one can determine all oscillations with small amplitudes, and can detect time intervals with significant amplitude of oscillations as well as their frequency variations.

Second, the wavelet spectrum of the AMO data reveals minor centennial Suess - de Vries cycles with periods from 195– to 235 – year; variable pieces of Gleisberg cycles with periods 70 – 130 years; a mode of solar rotation with period 50-60 years; and variable solar harmonics with periods 20 – 50 years, while of the TSI data shows almost constant Suess - de Vries cycles with a period from 195 – to 235 – year; variable Gleisberg cycles; and constant period of 11-year cycles for the first half of time series.

5.3 Wavelet coherence of TSI and AMO

In order to search for statistically relevant correlation between the periodicities found in the AMO and TSI data, we performed a cross wavelet and wavelet coherence analyses. The CWT is used to analyze the AMO and TSI signals. Similar to a common power spectrum, the CWT expands a time series into frequency space, but without losing the time information and allows analyzing time-localized oscillations. To find a relationship between the AMO and the TSI, we use wavelet coherence (WC) which compares two CWTs and finds locally phase-locked behavior of two signals. By using Monte Carlo simulations, the statistical significance level of the WC is tested. For each scale of the CWT, the significance level is estimated by calculating the wavelet coherence with each pair of datasets (Grinsted et al., 2004).

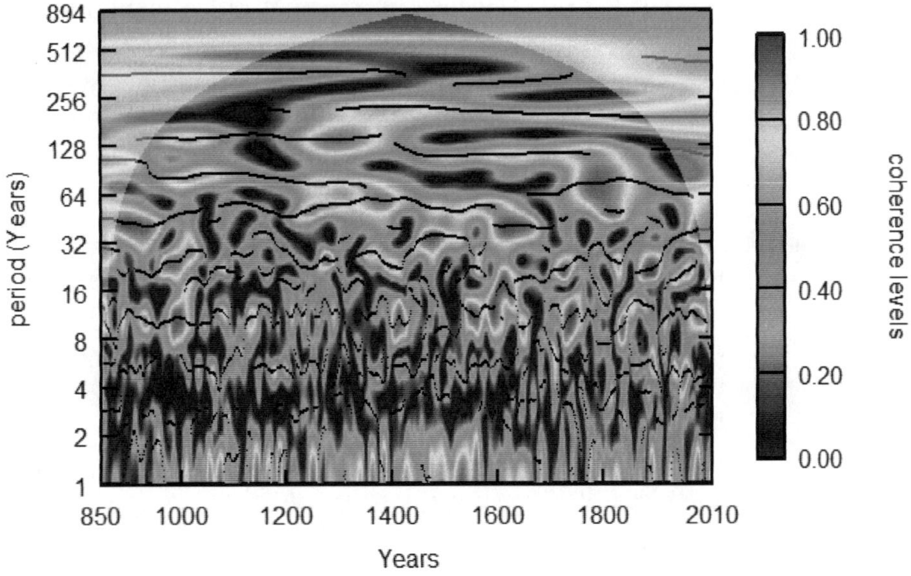

Figure 3: Wavelet coherence spectrum between the two TSI and AMO time series.

The relationship between the AMO and TSI annual time series over the period 850–2011 years was calculated by wavelet coherence. The absolute coefficients are shown in Fig. 3. As we known from (Chapanov, 2021) between the two TSI and AMO time series there are correlation periodicities shown in Table 1. In three dimensions, the results of wavelet coherence in fig.3 show significant coherence between the AMO and TSI. It is used to quantify the degree of association of two nonstationary series in the time – frequency domain.

6. CONCLUSIONS

This paper mainly focuses on the relationship between sunspot number and Atlantic multidecadal oscillation.

First, the results show that in contrast to the Fourier analysis, which determines all oscillations with small amplitudes, the CWT detects time intervals with significant amplitude of oscillations and their frequency variations.

Second, the obtained CWT of the AMO data reveals minor cycles; variable pieces of Gleisberg cycles; a mode of solar rotation; and variable solar harmonics, almost constant cycles for the TSI data and constant period of 11 – year cycles for the first half of time series.

Finally, the correlation between the AMO-TSI expressed by coherence is significant for oscillations close to 7-th harmonic of 2300-year Hallstatt solar cycle with 300 – year period; accelerated cycles of Suess - de Vries and accelerated Gleisberg cycles.

One important conclusion is that applications of both Fourier and wavelet analyses can significantly improve interdisciplinary research.

Acknowledgements

The study is supported by the National Science Fund of Bulgaria, Contract KP-06-N34/1 /30-09-2020 "Natural and anthropogenic factors of climate change – analyzes of global and local periodical components and long-term forecasts"

Data available

All the data used in this work are available freely in the internet. The TSI data used in this study are available online at *https://lasp.colorado.edu/lisird/data/nrl2_tsi_P1Y/* and the AMO data – at *https://www.ncei.noaa.gov/pub/data/paleo/reconstructions/wang2017/wang2017amv-amo.txt*

References

Benevolenskaya E. E., Kostuchenko I. G.: 2013, The Total Solar Irradiance, 2013. UV Emission and Magnetic Flux during the Last Solar Cycle Minimum, *Journal of Astrophysics,* vol. 2013, 368380.

Chapanov Ya.: 2020, *Variations of temperature over Bulgaria and their connection with solar cycles*, In: Proc. of the 1st International Conference on Environmental protection and Disaster Risks, Sofia, Bulgaria, pp. 22-33.

Chapanov Ya.: 2021, *Atlantic Multidecadal Oscillation Driven by Solar Harmonics*, In: K. Georgieva, B. Kirov and D. Danov (Edts.), Proc. of the Thirteenth Workshop "Solar Influences on the Magnetosphere, Ionosphere and Atmosphere", 183-188,

Cohen M.: 2019, A better way to define and describe Morlet wavelets for time-frequency analysis, *NeuroImage*, 199, 81-86.

Grinsted A., Moore J.C., Jevrejeva S.: 2004, Application of the Cross Wavelet Transform and Wavelet Coherence to Geophysical Time Series, *Nonlinear Processes in Geophysics*, 11, 561-566.

Goupillaud P., Grossmann A., Morlet J.: 1984, Cycle-octave and related transforms in seismic signal analysis, *Geoexploration*, 23(1), 85-102.

Fligge M., Solanki S. K.: 2000, The solar spectral irradiance since 1700, *Geophysical Research Letters*, 27(14), 2157-2160.

Foukal P.: 2012, A new look at solar irradiance variation, *Solar Physics*, 279(2), 365-381.

Fröhlich C., Lean J.: 1998, *Total Solar Irradiance Variations: The Construction of a Composite and its Comparison with Models*, Symposium - International Astronomical Union, 185, 89 – 102.

Lean J. L.: 2018, Estimating solar irradiance since 850 CE, *Earth Space Sci.*, 5, 133-149.

Mallat S.: 1999, *A Wavelet Tour of Signal Processing*, Academic Press.

Sadowsky J.: 1994, The continuous wavelet transform: a tool for signal investigation and understanding, John *Hopkins APL Tech. Digest*, 15(4), 306-318.

Sadowsky J.: 1996, Investigation of signal characteristics using the continuous wavelet transforms, *Johns Hopkins APL Technical Digest*, 17(3), 258-269.

Swingedouw D., Terray L., Cassou C., Voldoire A., Salas-Mélia D., Servonnat J.: 2010, Natural forcing of climate during the last millennium: Fingerprint of solar variability, *Clim. Dyn.*, 36(7-8), 1349-1364.

Shindell D. T., Schmidt G. A., Mann M. E., Rind D., Waple A.: 2001, Solar forcing of regional climate change during the Maunder minimum, *Science*, 294, 2149-2155.

Najmi S., Sadowsky J.: 1997, The continuous wavelet transform and variable resolution time-frequency analysis, *Johns Hopkins APL Tech. Digest,* 18(1), 134-140.

Enfield D. B., Mestas-Nunez A. M., Trimble P. J.: 2001, The Atlantic Multidecadal Oscillation and Its Relationship to Rainfall and River Flows in the Continental U.S., *Geophysical Research Letters,* 28, 2077-2080.

Velasco V. M., Mendoza B.: 2008, Assessing the relationship between solar activity and some large scale climatic phenomena, *Advances in Space Research*, 42, 866-878.

Wang J., Yang B., Ljungqvist F. C., Luterbacher J., Osborn T. J., Briffa K. R., Zorita E.: 2017, Internal and external forcing of multidecadal Atlantic climate variability over the past 1.200 years, *Nature Geosci.*, 10, 512-517.

Xu H., Lei B., Li Z.: 2021, A reconstruction of total solar irradiance based on wavelet analysis, *Earth and Space Science*, 8, e2021EA001819.

Proceedings of the XIII Bulgarian-Serbian Astronomical Conference
(XIII BSAC) Velingrad, Bulgaria, October 3-7, 2022
Editors: Evgeni Semkov, Milan S. Dimitrijević, Momchil Dechev and Zoran Simić
Publ. Astron. Soc. "Rudjer Bošković" No 25, 2023, 109-117

FOR OPTICAL FLICKERING IN SYMBIOTIC STAR MWC 560

DRAGOMIR MARCHEV[1], KIRIL STOYANOV[2], VLADISLAV MARCHEV[1,2], RADOSLAV ZAMANOV[2], BORISLAV BORISOV[1], DOROTEA VASILEVA[1], TEODORA ATANASOVA[1] and NATALIYA PAVLOVA[3]

[1]*Shumen University, Faculty of Natural Sciences, Department of Physics and Astronomy, 115 Universitetska Str., 9712, Shumen, Bulgaria,*
[2]*Institute of Astronomy and National Astronomical Observatory, Bulgarian Academy of Sciences, 72 Tsarigradsko Shose, 1784 Sofia, Bulgaria,*
[3]*Shumen University, Faculty of Mathematics and Informatics, Department of Algebra and Geometry, 115 Universitetska Str., 9712, Shumen, Bulgaria,*
E-mail: d.marchev@shu.bg, kstoyanov@astro.bas.bg, vmarchev@astro.bas.bg, rkz@astro.bas.bg, b.borisov@shu.bg, d.vasileva@shu.bg, t.atanasova@shu.bg, n.pavlova@shu.bg

Abstract. This study is based on observations of MWC560 during the last two observational seasons (2020/2021 and 2021/2022). Other than looking for flickering we were interested in following the variability of brightness in the same period. Looking for similarities in the spectra with other types of stars is also of great interest to us because it could help clarify the stellar configuration of such objects. Our observations during the last two observational seasons of MWC560 confirm the absence of flickering. From the similarities of the gathered spectra of XX Oph and MWC560 we assume that the components in XX Oph are a red giant and a white dwarf, which are also surrounded by a common shell.

1. INTRODUCTION

The symbiotic star MWC 560 (V694 Mon) is an object of high interest since 1990 when it exploded and the spectra showed mass ejection with velocities around 6000 km/s (Tomov et al. 1990). The untypical spectral behavior was accompanied by optical flickering in all colors. This flickering of MWC 560 was always present between 1984 and 2018 with a short exception during 2016. It completely disappeared in October 2018 (Goranskij et al. 2018). V694 Mon never returned in state of rest but entered a state of continuous brightness increase modulated by different periodicities of 331 and 1860 days and possible 9570 days (Munari et al. 2016).

This study is based on observations of MWC560 during the last two observational seasons (2020/2021 and 2021/2022). Other than looking for flickering we were interested in following the variability of brightness in the same period. Looking for similarities in the spectra with other types of stars is also of great interest to us because it could help clarify the stellar configuration of such objects.

2. OBSERVATION

In this study we present CCD photometry and spectral observations of MWC560. The photometry was done by the 40-cm telescope at the Astronomical observatory of the Shumen University (AOShU) (Kjurkchieva et al. 2020). The spectra were obtained with the ESPERO spectrograph (Bonev et al. 2017), mounted on the 2-m telescope at the Rozhen NAO.

The star was observed photometrically in two consecutive observational seasons during 2020-2022, a total of 7 nights were recorded. In Table 1. Is presented the journal of these observations, all of which are done with two Sloan filters g' and r'. Exposition was 10 s with g' and 5 s and 10 s with r'.

Table 1: Journal of observations of MWC 560 with the 40-cm telescope of the Shumen Astronomical Observatory.

Date	band	UT start – UT end hh:mm – hh:mm	N_{pts}	average [mag]	σ [mag]	amplitude [mag]
28 November 2020	g' r'	0:12 – 3:32	263 264	9.096 8.490	0.006 0.007	< 0.04
22 February 2021	g' r'	19:16 – 21:45	363 361	9.149 8.54	0.014 0.015	< 0.07 < 0.08
25 April 2021	g' r'	18:30 – 20:01	116 116	9.072 8.524	0.015 0.011	< 0.05 < 0.06
20 October 2021	g' r'	0:30 – 2:28	216 214	8.771 8.26	0.011 0.013	< 0.05 < 0.06
08 February 2022	g' r'	18:56 – 22:30	307 229	8.798 8.351	0.017 0.016	< 0.09 < 0.09
10 February 2022	g' r'	20:10 - 22:44 20:13 – 22:23	191 154	8.786 8.337	0.009 0.008	< 0.042 < 0.042
11 February 2022	g' r'	20:30 – 23:10	257 256	8.801 8.302	0.014 0.035	< 0.05 < 0.11

The FWHM of the target images were around 3 pix in g′and 4 pix in r′. The standard procedure was used for the reduction of the photometric data (de-biasing, dark frame subtraction and flat-fielding) using the software MaxImDL 6.2. We performed differential photometry using 4 standard stars. Their magnitudes (Table 2) were taken from the catalogue APASS DR9 (Henden et al. 2015). We transformed the data using equations 2 and 3 from article (Kjurkchieva et al. 2020).

Table 2: Photometrical standards used during the processing of the observations-MWC560

Label	Star ID	RA(2000)	Dec(2000)	g′	r′
C1	TYC 5396-491-1	07 26 00.69	-07 45 34.98	10.644	10.820
C2	TYC 5396-684-1	07 25 27.96	-07 44 41.84	9.013	7.913
C3	TYC 5396-916-1	07 25 43.28	-07 42 04.60	11.771	10.985
C4	TYC 5396-1090-1	07 25 58.36	-07 44 12.00	11.602	10.146
Chk	TYC 5396-1467-1	07 26 16.05	-07 50 10.14	10.315	9.360

In this work we presented and analyzed two spectra of MWC560 from 6.12.2019, when it was already in a high state and flickering was absent and of XX Oph from 2019 Jun 13. Both spectra are observed with the EspeRo Echelle spectrograph (Bonev et al. 2017) on the 2.0m telescope of the Rozhen National Astronomical Observatory, Bulgaria.

3. DISCUSSION

On 2020 November 28, for a 3.5 hours long observation run, we obtained 263 points in Sloan-g' and r' bands. During our run, no variability with amplitude larger then 0.04 mag is visible. The mean magnitudes are g'= 9.096 ± 0.006, r'= 8.490 ± 0.007. On Fig.1 is plotted the light curve together with observations from 2015 when flickering with amplitude 0.4 mag is visible. This CCD photometry of MWC 560 was performed with the 50/70 cm Schmidt telescope at the Rozhen National Astronomical Observatory.

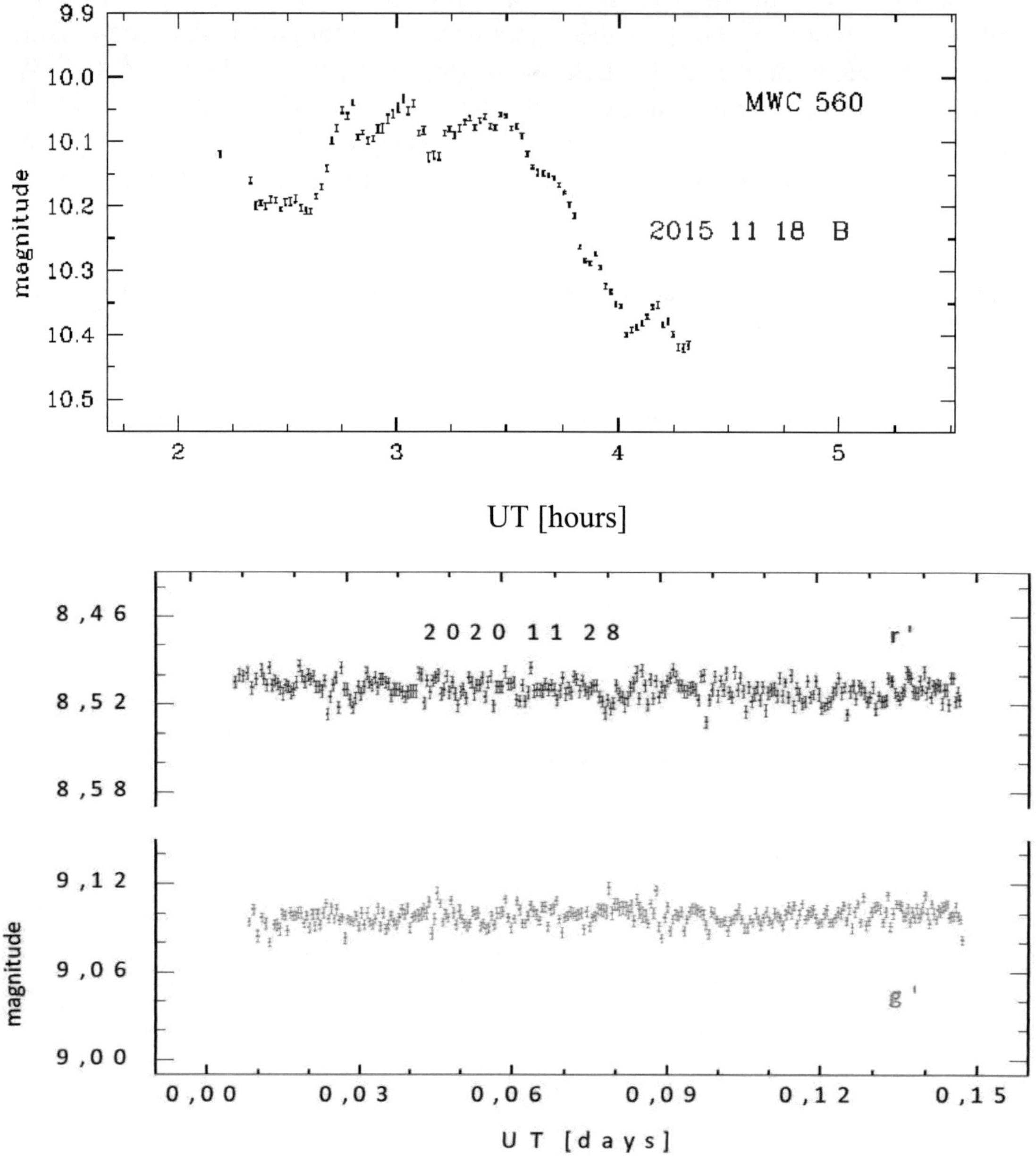

Figure 1: Observation MWC560 when flickering (2015, Rozhen NAO) and missing flickering (2020, AOShU).

Our first observations (2020 November 28) indicate that the optical flickering is still missing (Zamanov et al. 2020). Our observations in 2021 are from three nights presented in Fig.2. These observations also confirm the absence of flickering (Zamanov et al. 2021).

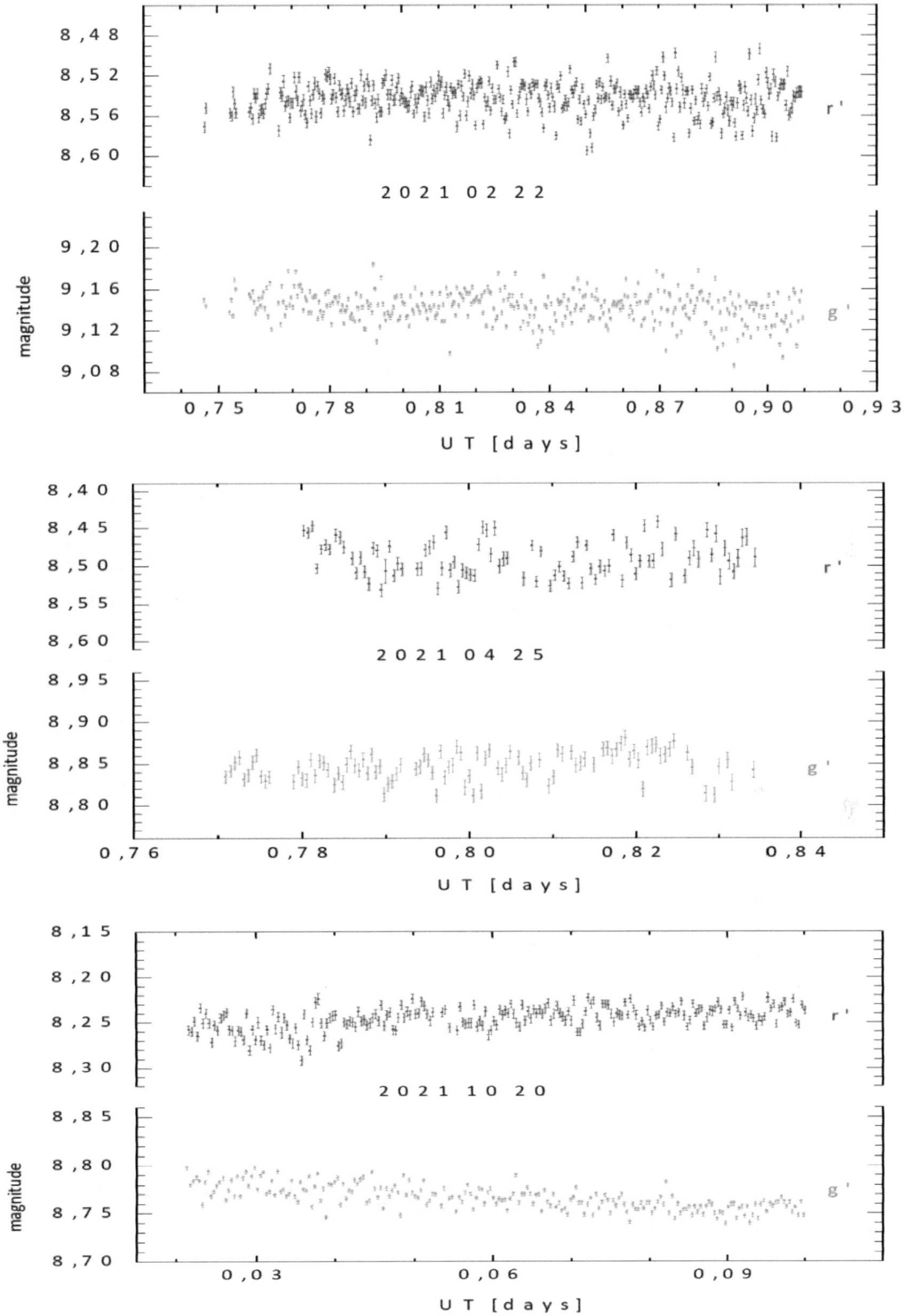

Figure 2: Observation MWC560 still missing flickering 2021.

In 2022 so far, we have observed the star 3 times in February and the results are shown in Fig. 3. There we do not see flickering in any of the 3 nights. We were also interested in monitoring the brightness during these consecutive nights. It turned out that the brightness remains relatively high and constant, considering the estimated uncertainty ranges, but a little smaller than the maximum at October 2021. This shows a continuous high state during 2022 in agreement with the findings of (Kondratyeva et al. 2021) for the period of 2018-2021.

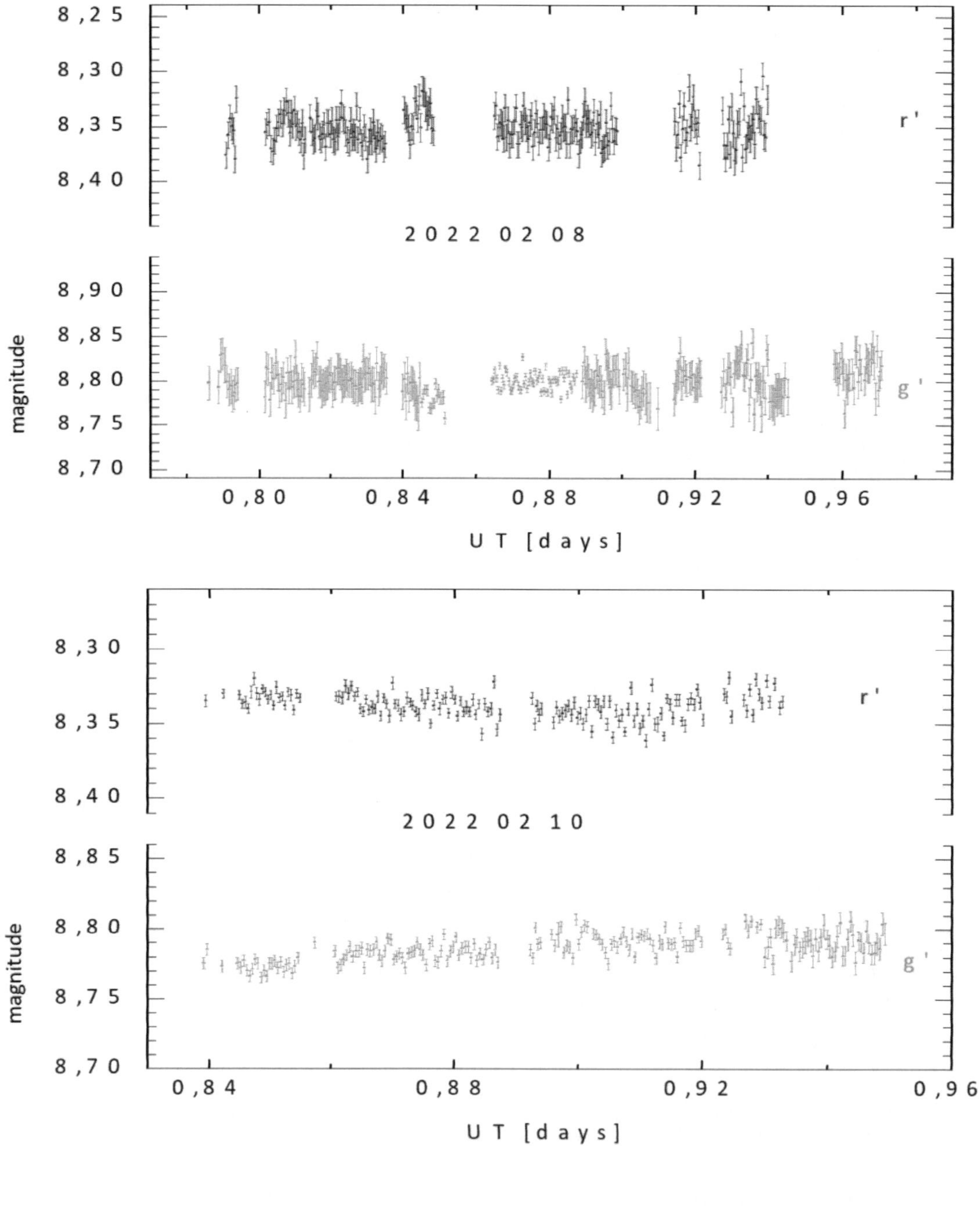

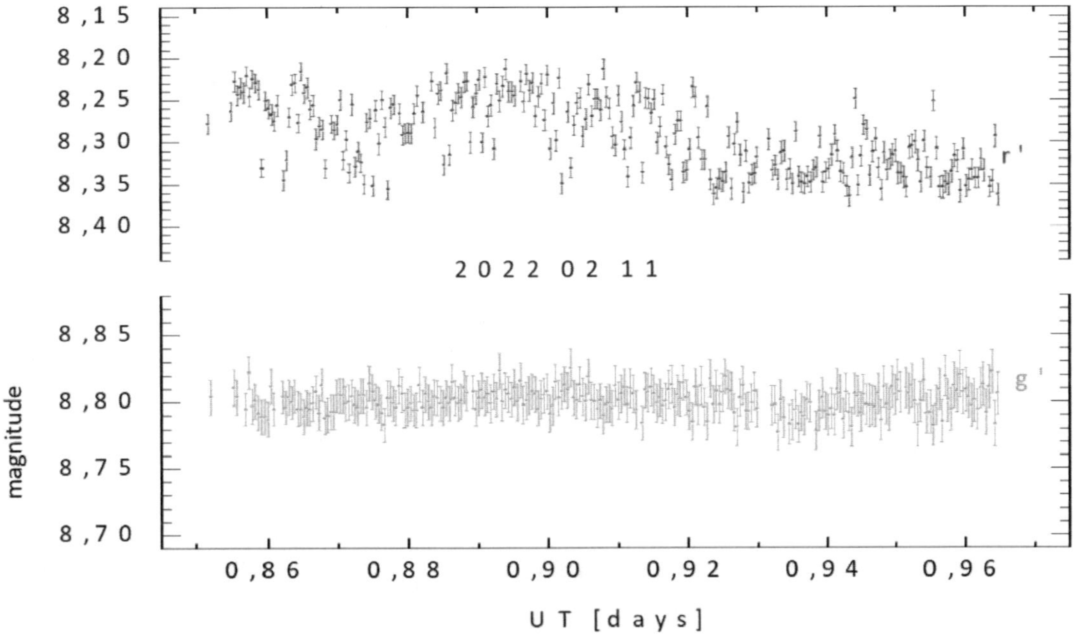

Figure 3: Observation MWC560 still missing flickering February 2022.

The increasing brightness from 2016 until 2020 shown in all spectral observations for that period suggests an expanding common shell of the system (Ando et al. 2021). The formation of such a shell and partial or complete destruction of the accretion disk was also suggested by (Kondratyeva et al. 2021).

A study of the brightness variability of MWC560, but in a short period of time- two months, was presented by (Munari 2021). He reports a peak of the brightness at 2021 October 27. The spectra gathered by (Goranskij et al. 2021), in November 2021 also show fast changes which is in agreement with the idea for the common shell.

Our observations show a maximum of the brightness at 2021 October 20 respectively g'=8.771±0.011 и r'=8.26±0.013, which is in agreement with the presented by (Munari 2021). Analyzing the photometry from February 2022 we can conclude that the system is still in a high brightness state, which most likely is a result of the powerful common shell.

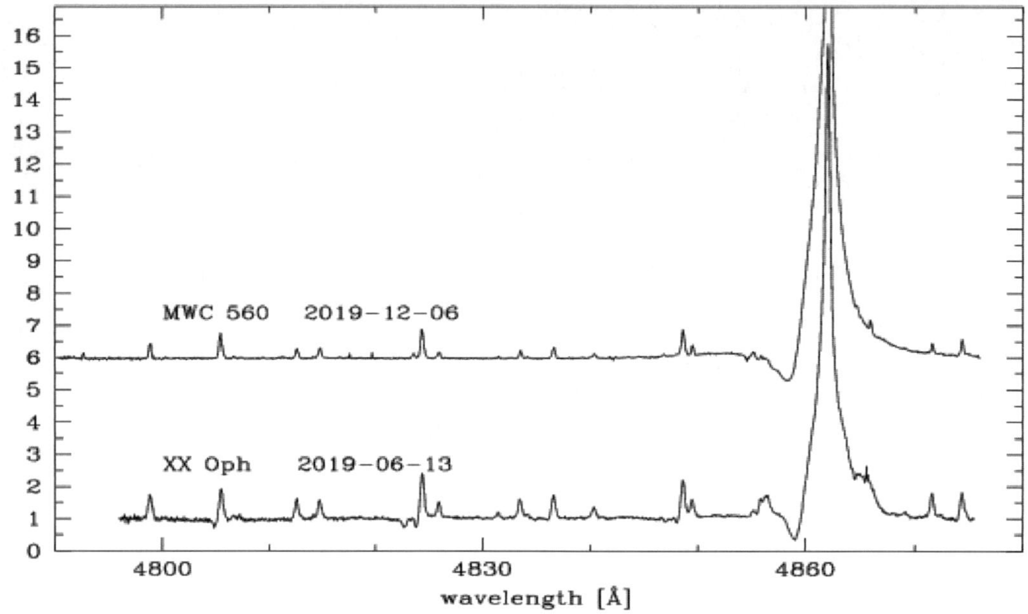

Figure 4: Similarity between the spectrum of MWC 560 (observed on 2019 December 6) and the spectrum of the Merrill's iron star XX Oph.

On Fig.4 are plotted the spectrum of MWC 560 obtained on 2019 December 6 and the spectrum of XX Oph obtained on 2019 June 13. The spectra are almost identical. At the time when this spectrum of MWC 560 is obtained, the flickering was missing and it was probably in stage of formation of a common envelope due to the transit of the system to a dynamical mode of accretion with an increased rate (Goranskij et al. 2018).

From the observations of MWC 560 during the last three decades, it is known that it consists of a mid-M non Mira giant and an white dwarf (Lucy et al. 2020). XX Oph is one of the two stars listed as "Iron Stars" (Bopp and Howell 1989) due to the appearance of metal emission lines in its optical spectrum. There are no clear evidences about the components of XX Oph (Howell et al. 2009). The spectral similarity shown in Fig.4 is a clue that likely the components inside XX Oph are a red giant and a white dwarf. The white dwarf is probably accreting at high mass accretion rate and the components are surrounded by envelope.

This similarity can be interpreted vice versa- in this stage of the evolution of MWC560, it is mimicking behavior of an "iron star".

4. CONCLUSIONS

Our observations during the last two observational season of MWC560 confirm the absence of flickering. The brightness of the system remains high, with a maximum in October 2021, also stated by (Munari 2021). The observations from February 2022 show a high level of brightness and absence of flickering. The absence of flickering and the increased brightness are most likely caused by a wide common shell of the system which is hiding the accretion disk of the compact object.

From the similarities of the gathered spectra of XX Oph and MWC560 we assume that the components in XX Oph are a red giant and a white dwarf, which are also surrounded by a common shell.

Acknowledgement

This article was partially funded by projects: RD-08-100/2022 and RD-08-146/2022 of the Scientific Research Fund of Shumen University and DO-01-383/2020 (RACIO) of the Scientific Research Fund of the Bulgarian Ministry of Education and Science.

References

Ando K., Fukuda N., Sato B., et al.: 2021, *PASJ*, 73, L37.
Bonev T., et al.: 2017, *Bulg. Astron. Journal,* 26, 67.
Bopp B. W., Howell S. B.: 1989, *PASP*, 101, 981.
Goranskij V. P., et al.: 2018, *ATel.*, #12227.
Goranskij V. P., et al.: 2021, *ATel.,* #15061.
Henden A. A., Levine S., Terrell D., Welch D. L.: 2015, *AAS*, 225, 336.
Howell S. B., et al.: 2009, *PASP*, 121, 16.
Kjurkchieva D., et. al.: 2020, *Bulg. Astron. Journal,* 32, 113.
Kondratyeva L. N., et. al.: 2021, *ApJ*, 64, 306.
Lucy A. V., et al.: 2020, *MNRAS,* 492, 3107.
Munari U., et al.: 2016, *New A.,* 49, 43.
Munari U., Dallaporta S.: 2021, *ATel.,* #15066.
Tomov T., et al.: 1990, *Nature,* 346, 637.
Zamanov R., et al.: 2020, *ATel.*, #14239.
Zamanov R., et al.: 2021, *ATel.*, #14988.

Proceedings of the XIII Bulgarian-Serbian Astronomical Conference
(XIII BSAC) Velingrad, Bulgaria, October 3-7, 2022
Editors: Evgeni Semkov, Milan S. Dimitrijević, Momchil Dechev and Zoran Simić
Publ. Astron. Soc. "Rudjer Bošković" No 25, 2023, 119-124

DARK MATTER HALOS IN GALAXY MERGERS

STANISLAV MILOŠEVIĆ

University of Belgrade, Faculty of Mathematics, Department of Astronomy
Studentski trg 16, 11158 Belgrade, Serbia
E-mail: stanislav.milosevic@matf.bg.ac.rs

Abstract. In this paper, we investigated the influence of dark matter in galactic halos on the dynamics of galaxies in merger events, using N-body simulations. In the standard cosmological picture, large massive galaxies are formed by mergers of smaller ones. These mergers are very important for galactic dynamics and evolution. The largest portion of the galactic mass is the dark matter halo. Dark matter has a very important role due to dynamical friction and the formation of observed substructures. Here we investigated how properties of spherical dark matter halos and merger circumstances influence the formation of these morphological structures.

1. INTRODUCTION

Galaxies are building blocks of the Universe. On large scales, we can observe the distribution of galaxies and calculate some cosmological parameters. In the local universe, we can investigate the properties of particular galaxies. In the bottom-up hierarchical model of galaxy formation, large galaxies are formed in mergers of smaller ones. In these mergers, many parameters are important: the mass of the galaxies, their rotation, gas content, and the timescale of the merger event. One of the most important parts of the galaxies is their dark matter halo. The largest portion of the mass of the galaxy is in the dark matter halo and for that reason, a halo is important for galactic dynamics and the formation of the structures in merger events.

There are several morphological types of galaxies. Hubble made the classification known as Hubble's diagram, or Hubble's tuning fork (Fig. 1).

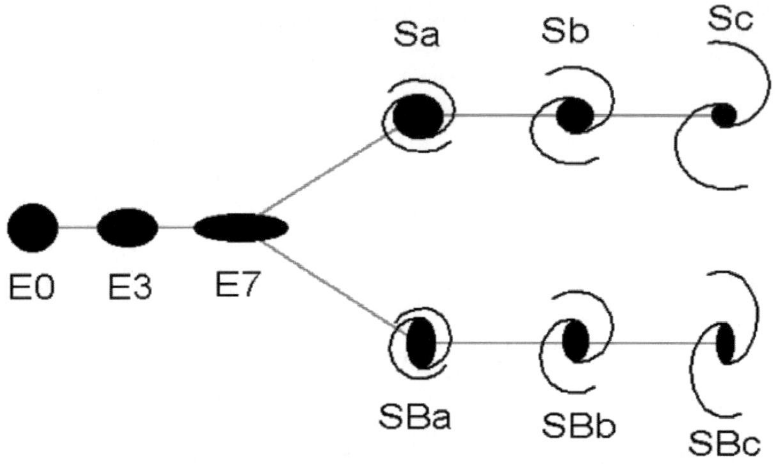

Figure 1: Hubble's "tuning fork" diagram.
(credit: The Ohio State University)

The main morphological types are elliptical galaxies, spiral and irregular galaxies. Elliptical galaxies have several types depending on the observing angle, their ellipticity changes with that angle. Spiral galaxies have disks and they are rotational systems. The velocity of rotation is larger than the velocity dispersion and we have an opposite situation with ellipticals. The large values of the rotational velocity or the flattening of the rotation curves are due to the present of the dark matter halo. Also, we have more gas in spirals then in ellipticals, and younger stars as a consequence. Irregular galaxies are mostly galaxies in interaction and they are tidally disrupted.

In this paper, we are interested in mergers of spiral galaxies and dwarf galaxies. Dwarf galaxies also could be spirals, ellipticals, and spheroidals. These mergers are forming structures in the halos of the spiral galaxies (Amorisco et al. 2013; Atkinson et al. 2015; Karademir et al. 2019) which we call the host and dwarf galaxy we will call progenitor. Examples of these mergers in the Local Group and structures formed in these dynamical processes are the Magellanic Stream in the halo of the Milky Way (Wannier, P., & Wrixon 1972) and Giant Stellar Stream (GSS) in the halo of the Andromeda galaxy. The discovery of the GSS is presented by Ibata et al. 2001. Many theoretical papers investigated the properties of these stellar streams, then shells and other structures, and these papers are based on N-body simulation.

The main influence of dark matter is the stability of the spiral galaxy, dynamical friction, and timescale of forming structures in mergers. In this paper, we present the main result of the N-body simulations of the merger event between the large spiral galaxy and dwarf elliptical galaxy (dSph) and investigate the role of the dark matter halo of the host in the merger event and galaxy dynamics. In Section 2 models of spiral and dwarf galaxies are presented; in Section 3 we present the main results, and in Section 4 we make a brief discussion and conclusions.

2. METHODS

We assume three components for the N-body model of the spiral galaxy: bulge, disk, and dark matter halo. For the model of the spherical bulge we took a density profile that is described with equation (Font et al. 2006; Fardal et al. 2007; Sadoun et al. 2014; Milošević et al. 2022):

$$\rho_b = \rho_0 (r/r_b) \exp(r/r_b)^{(-1/n)}.$$

The central density is ρ_0, and r_b is the bulge scale length. The density profile of the disk is modeled as two components profile. The exponential is in the radial direction:

$$\Sigma(R) = (M_d / 2\pi R_d^2) \exp(-R/R_d).$$

Here, M_d is the mass of the disk and R_d is the disk scale length in the radial direction. When we add the density profile in the z-direction which is sech^2, the density profile of the disk is given with:

$$\rho(R,z) = (\Sigma(R)/2z_0) \, sech^2(z/z_0),$$

where z_0 is the scale height of the disk. Dark matter halo is presented with Navarro-Frenk-White (NFW) profile (Navarro et al. 1996):

$$\rho_h(r) = \frac{\rho_0}{r/r_h (1 + r/r_h)^2},$$

where r_h is the scale radius of the dark matter halo. Values of the parameters are given in previous works cited above. The N-body model of the dwarf galaxy contains a spherical bulge and spherical dark matter halo presented with the same equations as for the spiral galaxy, but with different values of the parameters. To generate N-body models we used GalactICs package (Widrow, et al. 2008).

We run a large number of N-body simulations to find the influence of the orbits orientation on structure formation in the halo of the host and how,the dark matter halo influences the timescale of the merger event. Dark matter in the galaxy is very important for the dynamics, and the N-body representation of the halo is very important for calculating dynamical friction. For running simulations, we used Gadget2 cosmological code (Springel 2005).

3. RESULTS

In a galaxy merger between a large spiral host and a dSph galaxy, structures are formed in the halo of the host due to tidally disruption. In the first stage, the dark matter halo of the dSph galaxy starts to disrupt and baryonic particles are still bounded. These situations are presented in Fig. 3 and Fig. 4.

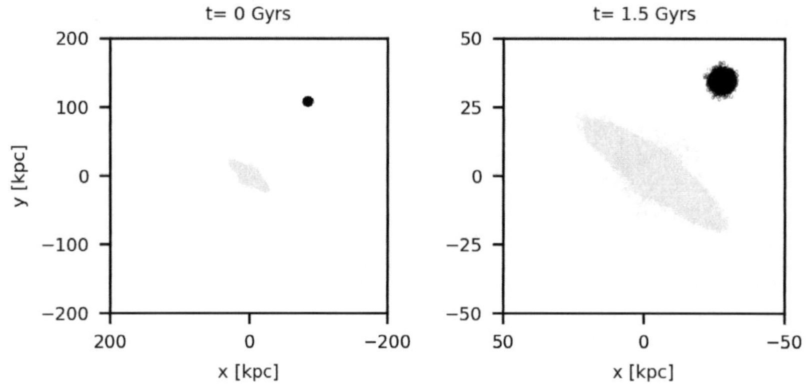

Figure 2: The baryonic matter particles of the dwarf galaxy are in black and the spiral galaxy is in yellow at the beginning of the simulation (left panel) and after 1.5 Gyrs (right panel). The baryonic part of the dSph galaxy isn't yet tidally disrupted.

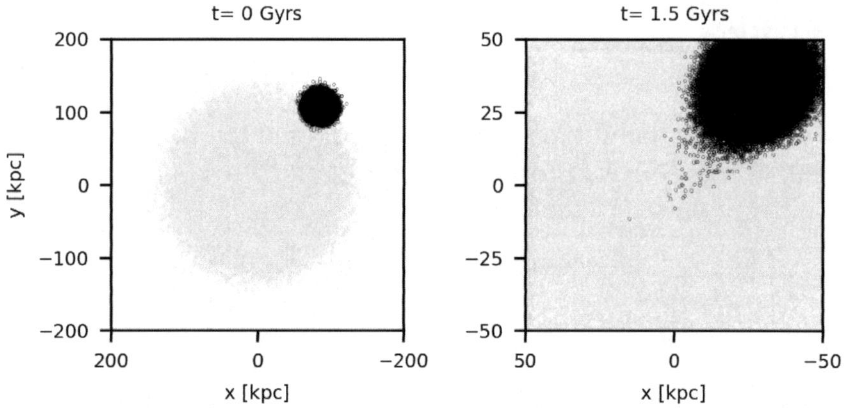

Figure 3: The dark matter halo particles of the dwarf galaxy are in black and the halo of the spiral galaxy is in yellow at the beginning of the simulation (left panel) and after 1.5 Gyrs (right panel). The dark matter halo of the dwarf is "pilling off" in the potential of the host.

Without the dark matter halo, the timescale would be different and the baryonic part of the dwarf would make structures in the halo of the host much earlier. Some of the formed structures are presented in Fig. 4:

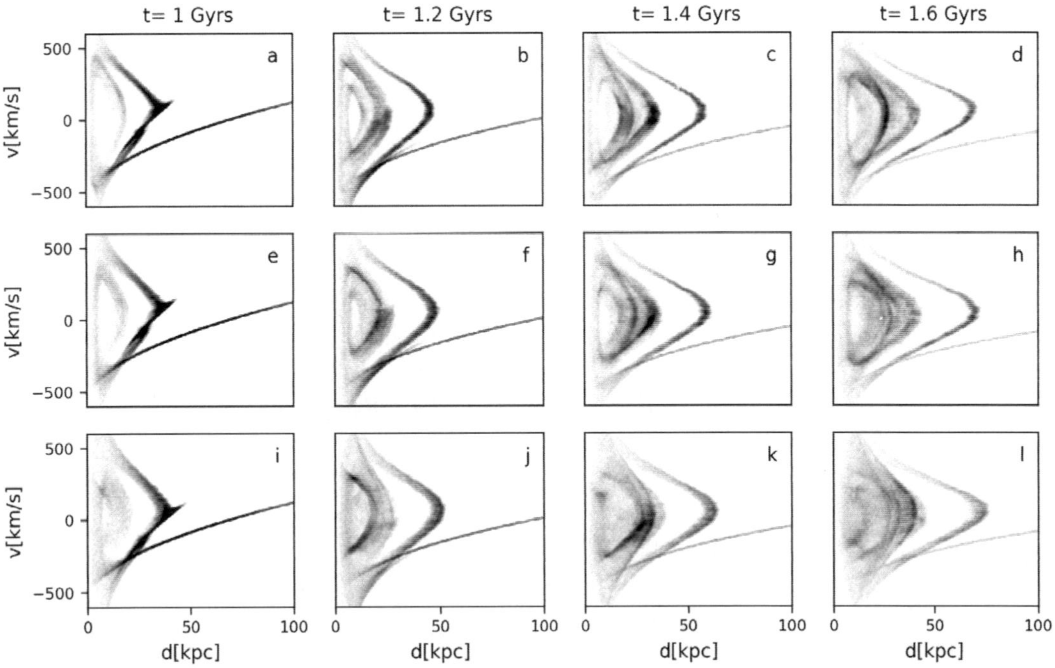

Figure 4: Structures in the halo of the host formed in a merger event, where progenitor is dSph for different infalls. We can see shells and stream, for the presented timescale which is between 1 and 1.6 Gyrs. This timescale of the formation of these shells and streams depends on the content of the dark matter. This graph is made as a continuation of research presented in Milošević 2022.

In Fig. 4 we can see formed structures in the halo of the host, from different infall angles of the dSph progenitor. These panels represent the baryonic matter of the dwarf in space d-v, where d is the distance from the center of the host, and v is radial velocity. In each simulation, we can see formed at least two stellar shells and one stellar stream. The properties of these structures are very dependent on the properties of the dwarf and the dark matter in the halo of the dwarf.

4. DISCUSSION AND CONCLUSIONS

We used pure N-body simulation to describe the dynamical history of galaxies. In particular, we were interested in the merger between a large spiral galaxy and its satellite, a dwarf spheroidal galaxy.

In merger events dark matter halo plays important role in the determination of the orbit of the dwarf in the potential of the host and the timescale of structure formation in the halo of the host.

Due to dynamical friction, the baryonic matter of the dwarf later starts to disrupt, because of the disruption of the halo of the dwarf.

Determination of the orbits of galaxies unlocks the possibility to determine the masses of the galaxies and makes constraints on the masses of the halos of both galaxies. In merger events, we found that stellar shells and streams are formed, and that is in agreement with previous works. In future work, it would be important to investigate gas dynamics in the mergers due to the important role that gas has in galaxy evolution.

References

Atkinson A. M., Abraham R. G., Ferguson A. M. N.: 2013, *ApJ*, 765, 28.
Amorisco N. C.: 2015, *MNRAS*, 450, 575.
Fardal M. A., Guhathakurta P., Babul A., McConnachie A. W.: 2007, *MNRAS*, 380, 15.
Geehan J. J., Fardal M. A., Babul A., Guhathakurta P.: 2006, *MNRAS*, 366, 996.
Ibata R., Irwin M., Lewis G., Ferguson A. M. N., Tanvir N.: 2001, *Natur,* 412, 49.
Karademir G. S., Remus R.-S., Burkert A., et al.: 2019, *MNRAS*, 487, 318.
Milošević S.: 2022, *Serbian Astronomical Journal.*
Milošević S., Mićić M., Lewis G. F.: 2022, *MNRAS*, 511, 2868.
Navarro J. F., Frenk C. S., White S. D. M.: 1996, *ApJ*, 462, 563.
Sadoun R., Mohayaee R., Colin J.: 2014, *MNRAS*, 442, 160.
Springel V.: 2005, *MNRAS*, 364, 1105.
Wannier P., Wrixon G.T.: 1972, *ApJ*, 173, 11.
Widrow L. M., Pym B., Dubinski J.: 2008, *ApJ*, 679, 1239.

Proceedings of the XIII Bulgarian-Serbian Astronomical Conference (XIII BSAC) Velingrad, Bulgaria, October 3-7, 2022
Editors: Evgeni Semkov, Milan S. Dimitrijević, Momchil Dechev and Zoran Simić
Publ. Astron. Soc. "Rudjer Bošković" No 25, 2023, 125-135

GEOMAGNETIC STORMS AND THEIR SOLAR ORIGIN IN SOLAR CYCLE 24 (2009 – 2019)

ROSITSA MITEVA[1], SUSAN W. SAMWEL[2] and MOHAMED NEDAL[1]

[1]*Institute of Astronomy with National Astronomical Observatory - Bulgarian Academy of Sciences, 72 Tsarigradsko Chaussee Blvd., 1784 Sofia, Bulgaria*
[2]*National Research Institute of Astronomy and Geophysics (NRIAG), 11421, Helwan, Cairo, Egypt*
E-mail: rmiteva@nao-rozhen.org, rmiteva@astro.bas.bg, samwelsw@nriag.sci.eg, mnedal@astro.bas.bg

Abstract. We present a list of 171 geomagnetic storms with Dst index ≤ −50 nT that occurred in the solar cycle 24 (2009−2019) as currently reported by the Kyoto database. Solar origin association is proposed in terms of coronal mass ejections (CMEs), together with the accompanying interplanetary CMEs and shocks close to Earth. Correlation between the strength of the Dst index and the speeds of the solar and heliospheric phenomena is performed and the main results are discussed.

1. INTRODUCTION

Geomagnetic storms (GSs) are disturbances in the magnetosphere of the Earth (Dungey, 1961) which can be detected by a network of magnetometers at the ground level. Different combinations of stations are used to compile various geomagnetic indices. The one used to reflect the strength of the GSs is the disturbance storm time (Dst), measured in negative values, in nT, and used in the present work. It is widely accepted the origin of the GSs to be the interplanetary (IP) counterparts of the coronal mass ejections (CMEs) – ICMEs, or alternatively, fast solar wind streams (Richardson, 2018). CMEs are the magnetized plasma, ejected from the corona due to some large scale reconfiguration, and propagating outwards in the IP medium (Webb and Howard, 2012). Upon the ICME arrival at Earth, shock waves may form, usually detected by in situ measurements. GSs are key players in solar and heliospheric studies. They perturb not only the Earth's atmospheric layers, but may also cause a negative effect on ground-based and satellite technologies and indirectly to humans (these effects are termed as space weather, Temmer, 2021). Thus various forecasting efforts are proposed in the search of practical mitigation strategies (Balan et al., 2019).

The aim of this study is to compile a list of all GSs with Dst index ≤ 50 nT in solar cycle (SC) 24 and to identify the associated CMEs, ICMEs and IP shocks, if possible.

2. DATA

Hourly reports from the World Data Center (WDC) for Geomagnetism, Kyoto, were used for this study. The final version of the reports from this database (https://wdc.kugi.kyoto-u.ac.jp/dst_final/) are available by the end of 2016. For the remaining three years in our list we used the provisional reports (https://wdc.kugi.kyoto-u.ac.jp/dst_provisional/index.html), which are however subject to change.

The CME parameters were collected from the SOHO LASCO CME catalog: https://cdaw.gsfc.nasa.gov/CME_list/

The ICME speed is taken from the online version of the Richardson and Cane list: https://izw1.caltech.edu/ACE/ASC/DATA/level3/icmetable2.htm If reported, we take into consideration the CME identified as the solar origin of the ICME.

IP shock speed is taken from the shock database (available at present only by 2018): http://www.ipshocks.fi/database We select IP shocks (from Wind spacecraft) only in about a one-day period prior to the hour of the minimum Dst (regarded as the GS peak time). In case of multiple shocks in close succession that are reported in this time, we list only the earlier entry of the IP shock speed. In the online version of the presented here GS list (https://catalogs.astro.bas.bg/) we will present the time and hour of the shock at the Wind spacecraft and the speeds calculated at both Wind and ACE satellites. For the purpose of this report, we do not distinguish between fast forward and fast reverse type of the shocks.

3. SOLAR ORIGIN IDENTIFICATION

The association procedure that was followed is presented below. We search for the parent CME giving rise to the GS in a 3 (to about 5) day period prior to the hour of the minimum of the Dst index, as reported. In this time window, we select the fastest and widest CME which is preferably Earth-directed (observed as halo, i.e. 360 degrees in angular width (AW), or partial halo), as the most probable origin of the GS. This procedure, however, cannot differentiate between forward and backward propagations. Thus we use ready to inspect ENLIL animations (http://helioweather.net/) in order to confirm the directivity of expansion of the solar wind or/and CME structures and finally to select a parent CME.

In summary, we could identify 70/171 (41 %) CMEs as the parent of the GSs in SC24, as listed in Table 1, with the majority (48/70, 69 %) being halo CMEs. Furthermore, we found 70/171 (41 %) ICMEs and 67/171 (39 %) IP shocks related to the GSs. We note that there is unavoidable subjectivity in the procedure of solar origin and related phenomena association. The GSs and their proposed associations are listed in Table 1.

4. RESULTS

Table 1: List of GSs in SC24, their parent CME, ICME and IP shock. All times are in UT, speeds are in km/s, AW in degrees, Dst index in nT, u - uncertain.

GS date & time yyyy-mm-dd-hr	|Dst|	CME day/time/speed/AW	ICME speed	IP shock speed
2009-07-21-07	83	u	330	u
2010-02-15-24	59	u	no	338
2010-04-06-15	81	03/10:34/668/360	640	838
2010-04-12-02	67	08/04:54/264/160	410	444
2010-05-02-19	71	u	no	no
2010-05-29-13	80	24/14:06/427/360	360	297
2010-06-04-02	53	31/21:08/406/115	no	no
2010-08-04-02	74	01/u/850/360	530	531
2010-10-11-20	75	07/07:24/417/111	no	no
2011-02-04-22	63	01/23:24/437/360	430	363
2011-03-01-15	88	24/07:48/1186/158	no	no
2011-03-11-06	83	07/20:00/2125/360	no	no
2011-04-06-19	60	04/15:57/2081/109	no	no
2011-05-28-12	80	u	510	no
2011-07-05-01	59	01/13:36/356/112	no	no
2011-08-06-04	115	04/04:12/1315/360	540	789
2011-09-10-05	75	06/23:06/575/360	470	340
2011-09-17-16	72	13/24:00/408/242	430	509
2011-09-26-24	118	24/12:48/1915/360	580	544
2011-10-25-02	147	22/10:24/1005/360	460	542
2011-11-01-16	66	27/12:00/570/360	380	no

2012-01-23-06	71	19/14:36/1120/360	450	443
2012-01-25-11	75	23/04:00/2175/360	no	630
2012-02-15-17	67	09/21:18/659/360	no	no
2012-02-19-05	63	u	no	no
2012-02-27-20	57	24/03:46/800/189	440	no
2012-03-02-02	54	u	no	no
2012-03-04-02	50	u	390	no
2012-03-07-10	88	04/11:00/1306/360	no	485
2012-03-09-09	145	07/00:24/2684/360	550	1245
2012-03-12-17	64	09/04:26/950/360	no	520
2012-03-15-21	88	13/17:36/1884/360	680	711
2012-03-28-05	68	24/00:24/1152/360	no	no
2012-04-05-08	64	02/02:12/350/135	no	no
2012-04-13-05	60	u	no	u
2012-04-24-05	120	u	370	416
2012-06-12-02	67	08/19:48/371/120	no	no
2012-06-17-14	86	14/14:12/987/360	440	486
2012-07-09-13	78	04/17:24/662/360	410	no
2012-07-15-17	139	12/16:48/885/360	490	617
2012-09-03-11	69	01/20:00/1442/360	310	u
2012-09-05-06	64	02/03:09/538/360	500	399
2012-10-01-05	122	29/00:12/947/360	370	452
2012-10-09-09	109	05/02:48/612/284	390	445
2012-10-13-08	90	09/00:48/692/122	490	no
2012-11-02-21	65	27/16:48/317/360	340	391

2012-11-14-08	108	09/15:12/559/276	380	386
2013-01-17-24	52	u	390	no
2013-01-26-23	51	23/14:12/530/185	no	343
2013-03-01-11	55	u	no	no
2013-03-17-21	132	15/07:12/1063/360	520	719
2013-03-29-17	59	u	no	no
2013-05-01-19	72	u	430	423
2013-05-18-05	61	15/01:48/1366/360	no	445
2013-05-25-07	59	22/13:26/1466/360	no	527
2013-06-01-09	124	u	no	410
2013-06-07-06	78	u	430	no
2013-06-29-07	102	28/02:00/1037/360	390	466
2013-07-6-19	87	01/20:24/819/208	350	no
2013-07-10-22	56	u	no	507
2013-07-14-23	81	09/15:12/449/360	430	no
2013-08-05-03	50	u	no	no
2013-08-27-22	59	u	no	no
2013-10-02-08	72	29/22:12/1179/360	470	515
2013-10-09-02	69	06/14:43/567/360	480	264
2013-10-30-24	54	28/15:36/812/360	no	343
2013-11-07-13	50	u	no	no
2013-11-09-09	80	u	420	u
2013-11-11-08	68	u	no	no
2013-12-08-09	66	u	no	369
2014-02-19-09	119	16/10:00/634/360	520	597

2014-02-20-07	80	u	no	821
2014-02-22-02	64	18/01:36/779/360	490	no
2014-02-23-20	55	20/03:12/993/360	no	no
2014-02-27-24	97	25/01:26/2147/360	no	438
2014-03-01-09	52	u	no	no
2014-04-12-10	87	u	350	no
2014-04-30-10	67	u	310	no
2014-08-27-19	79	u	no	no
2014-09-12-24	88	10/18:00/1267/360	600	747
2014-10-09-08	51	u	no	no
2014-10-28-02	65	u	no	no
2014-11-10-18	65	07/18:08/795/293	480	no
2014-11-16-08	59	u	no	no
2014-12-12-17	53	u	no	no
2014-12-22-06	71	u	380	475
2014-12-23-23	57	19/02:00/126/505	no	469
2014-12-24-24	53	u	no	no
2014-12-26-02	57	u	no	no
2015-01-04-22	78	u	no	no
2015-01-07-12	107	u	450	494
2015-02-02-07	55	u	no	389
2015-02-18-01	64	u	no	no
2015-02-24-08	58	u	no	no
2015-03-02-09	64	u	no	no
2015-03-17-23	234	15/01:48/719/360	560	562

2015-04-10-05	60	u	no	409
2015-04-11-10	85	u	380	no
2015-04-16-24	88	u	no	398
2015-05-13-07	82	u	no	u
2015-06-08-09	67	u	no	u
2015-06-23-05	198	21/02:36/1366/360	610	767
2015-06-25-17	81	22/18:36/1209/360	550	720
2015-07-05-06	74	u	no	no
2015-07-13-16	68	u	490	532
2015-07-23-09	72	u	no	no
2015-08-16-18	98	12/14:48/647/204	500	456
2015-08-19-07	64	u	no	no
2015-08-23-09	57	u	no	no
2015-08-27-21	103	22/07:12/547/360	370	no
2015-09-04-07	50	u	no	u
2015-09-07-21	75	u	no	no
2015-09-09-13	105	u	460	no
2015-09-11-15	87	u	no	no
2015-09-20-12	79	18/05:00/823/131	510	608
2015-10-04-10	63	u	no	no
2015-10-07-23	130	u	no	no
2015-10-18-10	56	u	no	no
2015-11-04-13	56	u	640	627
2015-11-07-07	87	04/14:48/578/360	500	943
2015-11-09-17	55	u	no	no

2015-11-10-14	56	u	no	no
2015-12-13-20	55	u	no	no
2015-12-20-23	166	16/09:36/579/360	400	563
2016-01-01-01	116	28/12:12/1212/360	440	404
2016-01-20-17	101	14/23:24/191/360	370	u
2016-02-01-09	53	u	no	no
2016-02-03-03	57	u	no	no
2016-02-16-20	65	u	no	no
2016-02-18-01	62	u	no	no
2016-03-06-22	99	u	no	no
2016-03-15-08	50	u	no	405
2016-03-16-24	56	u	no	no
2016-04-02-24	59	u	no	no
2016-04-07-22	61	u	no	no
2016-04-13-02	56	u	no	366
2016-04-14-21	61	u	420	558
2016-04-16-22	58	u	no	no
2016-05-08-08	95	u	no	no
2016-08-03-16	52	u	no	361
2016-08-23-22	73	u	no	no
2016-09-01-10	57	u	no	no
2016-09-02-03	59	u	no	no
2016-09-28-18	51	u	no	no
2016-09-29-10	65	u	no	no
2016-10-04-10	50	u	no	no

2016-10-13-18	110	u	390	431
2016-10-25-18	65	u	no	no
2016-10-29-04	70	u	no	no
2016-11-10-18	60	05/04:36/403/22	360	344
2017-01-01-22	61	u	no	271
2017-01-27-15	70	u	no	133
2017-04-22-17	51	u	no	no
2017-05-28-08	125	23/05:00/259/243	360	378
2017-07-16-16	72	14/01:26/1200/360	520	no
2017-08-31-12	51	u	no	no
2017-09-08-02	-122	04/20:36/1418/360	590	no
2017-09-28-07	56	u	no	no
2017-10-14-06	53	u	no	no
2017-11-08-02	73	u	no	no
2018-03-18-22	50	u	no	no
2018-04-20-10	60	u	no	no
2018-05-06-03	57	u	no	no
2018-08-26-07	175	20/21:24/126/120	410	u
2018-09-11-11	60	u	no	u
2018-10-07-22	53	u	no	u
2018-11-05-06	53	u	no	u
2019-05-11-22	51	u	350	u
2019-05-14-08	65	u	470	u
2019-08-05-21	53	u	no	u
2019-09-01-07	52	u	no	u

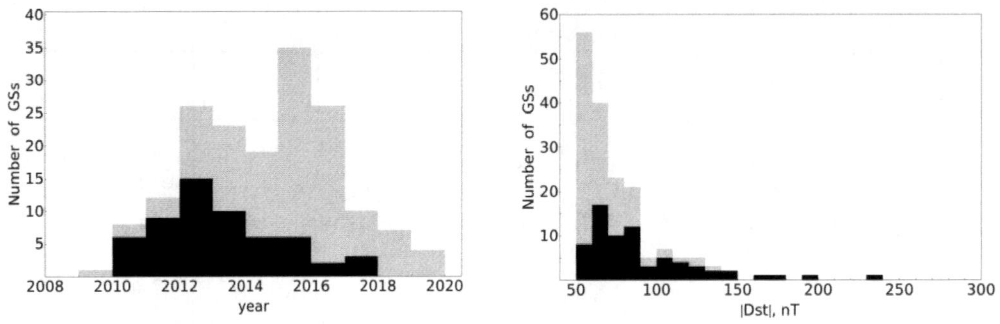

Figure 1: Stacked distribution of the GSs with years (left) and |Dst| (right). Black/Gray color is used for the GSs with/without identified solar origin.

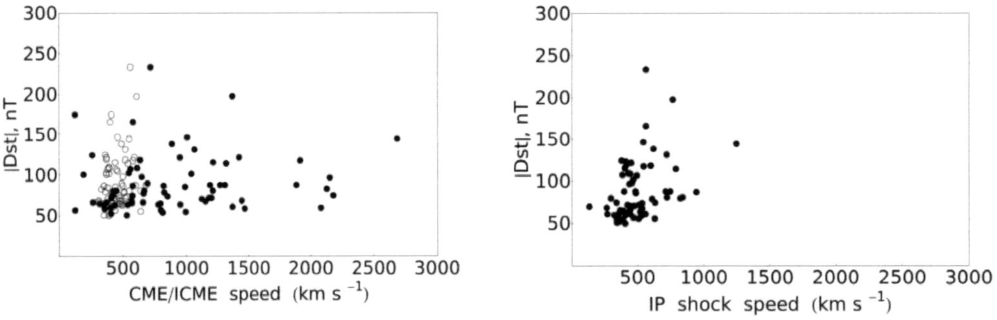

Figure 2: Scatterplots between the Dst index with the CME (with filled) and ICME (empty symbols) speed (left) or with the IP shock speed (right).

In Figure 1 are presented the distributions of the GSs in time (on the left) and of the module of the Dst index (on the right). The length of each color bar reflects the number of cases with (denoted with black color) or lack of (gray color) solar origin (CME) identification. The annual distribution of the GSs follows the sunspot trend. More CME-related GSs are detected in the first half of the SC24. Moreover, nearly all strong GSs (< -100 nT) can be traced back to solar eruptive phenomena, whereas for the majority of the weak GSs (145/171), defined here with Dst index in the range from -50 to -100 nT, it is more difficult to identify with certainty a solar origin (only 50/145, or 34% of them have identified CME).

Figure 2 presents the scatter plots between the strength of the GSs, measured in |Dst| and different speeds of the possible drivers, CME/ICME (left) and IP shock (right plot). For comparative purposes, the plots cover the same ranges. Overall, a large spread is seen with no clear-cut trends. For the ICME speed, we note that the data points are clustered over a very narrow velocity range.

5. CONCLUSIONS

We present a list of 171 GSs with Dst index ≤ −50 nT that occurred in SC24 (2009−2019) together with our proposition for their parent solar identifications and related phenomena. This constitutes a comprehensive scan of the Kyoto data compared to many available partial lists or lists of extreme GSs (e.g., Vennerstron et al. 2016). This catalog is supplemented by the ICMEs and IP shock speeds, as recorded near Earth. In total, 41% of all GSs could be associated with CME at the Sun, where the association decreases for weaker GSs due to the larger amount of uncertainties related to faint events. The parameters of the parent CMEs, together with the ICME and IP shock speeds are also provided. The entire list will be made available for scientific and operational purposes to the solar and space weather communities via the open access platform: https://catalogs.astro.bas.bg/

Acknowledgements

This work is supported by the Bulgarian-Egyptian interacademy project, IC-EG/08/2022-2024 and ASRT-BAS/2022-2023/10116.

References

Dungey J.W.: 1961, Interplanetary magnetic field and the auroral zones, *Phys, Res. Lett*, 6, 47-48.

Richardson I.G.: 2018, Solar wind stream interaction regions throughout the heliosphere, *Liv. Rev. Sol. Phys.*, 15, 1.

Webb D.F., Howard T.A.: 2012, Coronal mass ejections: Observations, *Liv. Rev. Sol. Phys.*, 9, 3.

Temmer M.:. 2021, Space weather: The solar perspective, *Liv. Rev. Sol. Phys.*, 18, 4.

Balan N., Zhang Q.H., Xing A. et al.: 2019, Capability of geomagnetic storm parameters to identify severe space weather, *ApJ*, 887, 51(1-11).

Vennerstrom S., Lefevre L., Dumbovic M. et al.: 2016, Extreme geomagnetic storms - 1868-2010, *Sol. Phys.*, 291, 1447-1481.

Proceedings of the XIII Bulgarian-Serbian Astronomical Conference (XIII BSAC) Velingrad, Bulgaria, October 3-7, 2022
Editors: Evgeni Semkov, Milan S. Dimitrijević, Momchil Dechev and Zoran Simić
Publ. Astron. Soc. "Rudjer Bošković" No 25, 2023, 137-142

PROCESSING DOUBLE STARS IMAGES USING MACHINE LEARNING

RADE PAVLOVIĆ[1], ALEKSA JOVANOVIĆ[2], ZORICA CVETKOVIĆ[1] and OLIVER VINCE[1]

[1]*Astronomical Observatory, Volgina 7, 11060 Belgrade, Serbia*
[2]*Faculty of Mathematics, University of Belgrade, Serbia*
E-mail: rpavlovic@aob.rs, ai18272@alas.matf.bg.ac.rs, zorica@aob.rs, ovince@aob.rs

Abstract. We use Machine Learning to process observations of double stars obtained by lucky imaging technique. First, we generate a large number of images of a double star system, whose PSF is Gaussian in the shape distribution, in a wider range of separations and position angles, which serves us for learning the system. Once we have trained the system to recognize double stars, we input our images of a particular double star, which we remade at the Astronomical Station at Vidojevica with the 1.4 m telescope attached CCD camera Andor iXon 897 Ultra, and as a result we get its separation and positional angle.

1. INTRODUCTION

Since CCD cameras were used to capture two stars at the end of the 20th century, the most common method for determining relative coordinates, separation and position angle, of double stars is the calculation of angular distance between the photocenters of the components and the angle between the lines which passes through the photocenters of components and direction to the north celestial pole calculated over the east. Here we suggest alternative method - application of machine learning i.e., convolutional neural network (CNN) for determining these double star parameters.

2. SIMULATED IMAGES OF DOUBLE STARS

To learn the system, we simulated images of binary stars. Knowing that our test star WDS 16238+6142 = STF 2054AB has a separation of about 1 arcsecond and a position angle of about 270 degrees we generated images in the range 0.7 to

1.3 arcsec with step 0.01 arcsec for separation and in the range of 250 to 290 degrees with step of 1 degree for positional angle[1]. Star profiles were 2D Gaussian

$$f_i(x,y) = A_i e^{-\left[\frac{(x-x_{0i})^2}{2\sigma_{Xi}^2}+\frac{(y-y_{0i})^2}{2\sigma_{Yi}^2}\right]}, i = 1,2$$

where A_i is the amplitude, x_{0i} and y_{0i} are the coordinates of the Gaussian center, σ_{Xi} and σ_{Yi} are variances for the given component axes. Labels are primary $i = 1$, and secondary component $i = 2$. In order to make the simulated images as realistic as possible, we added noise of the form Gaussian distribution and amplitude equal to the noise amplitude on the recorded images. The primary star is placed in the center images. The image resolution was 256 × 256 pixels. This resolution of simulated recordings was chosen to match the resolution of real images obtained on the telescope. As an example of a simulated image of double star is given in Fig. 1.

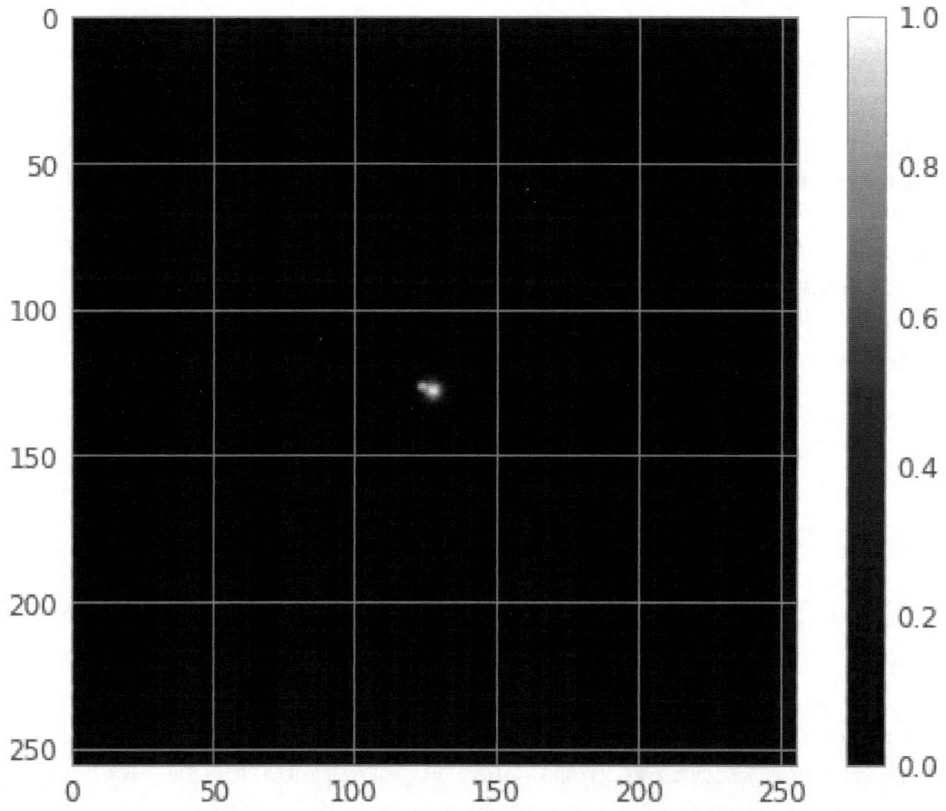

Figure 1: Example of simulated image with parameters $\rho = 0.70$ arcsec and $\vartheta = 250^o$.

[1] In our case, we measured positional angle in the local coordinate system related to the image plane.

3. REAL DOUBLE STAR IMAGES

Our equipment for lucky imaging mounted at the right port of "Milanković" telescope consists of two CCD cameras Andor iXon 897 Ultra and SBIG STXL-6303e and an Optec Perseus 4-port Instrument Selector as shown on Fig. 2. The detailed description of double star imaging procedure is given in Pavlović et al. (2021). As a test object we chose double star WDS 16238+6142 = STF 2054AB.

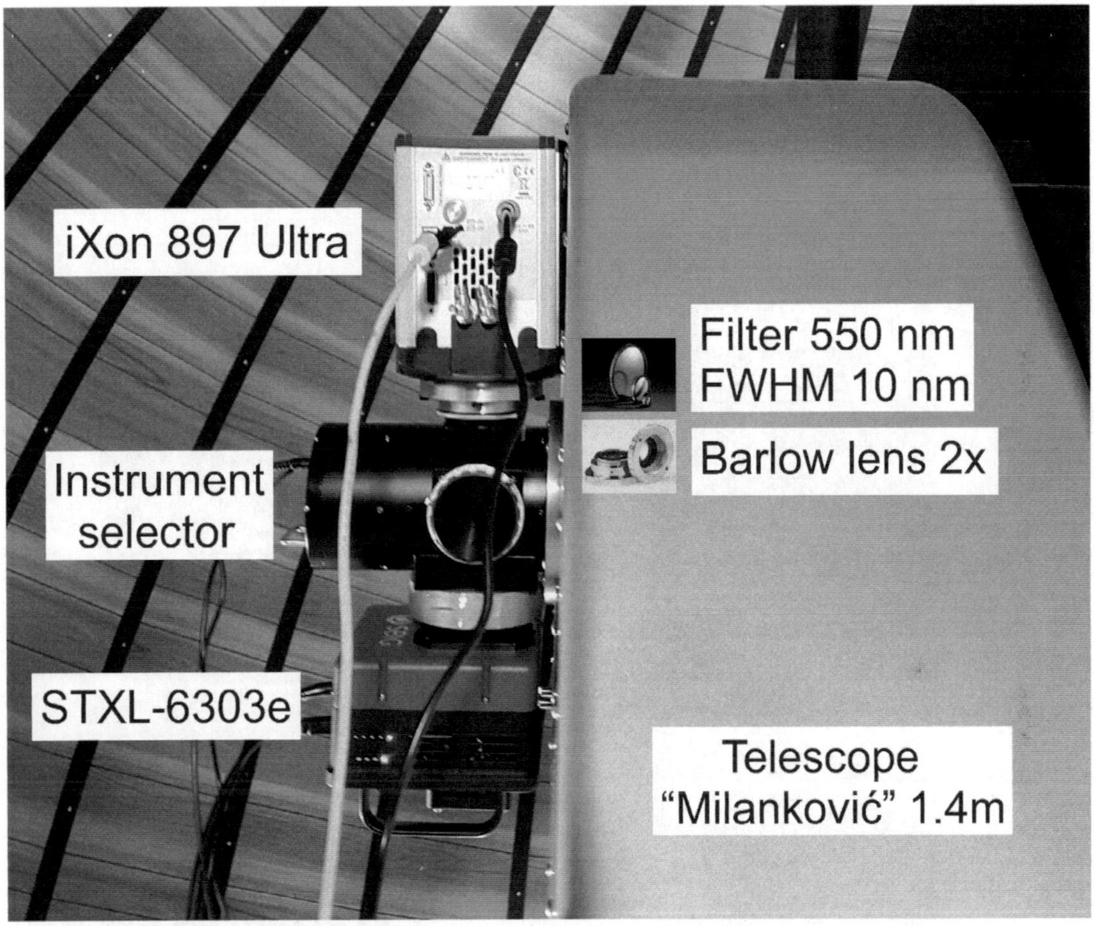

Figure 2: The equipment for lucky imaging attached to telescope "Milanković" at AS Vidojevica.

We recorded it twice using lucky imaging on September 16, 2021. The first set of recordings was made at 23:40:29 and consisted of 100 exposures, and the second one recorded at 23:42:24 and consisted of 1000 exposures. All images were acquired with an exposure of 50 ms. From both sets we chose 5% and 10% of the best frames. In that manner we have prepared four files to which should be applied convolutional neural network. Fig. 3 shows all four acquired frames of the double stars WDS 16238+6142 = STF 2054AB in reduced size as the details would be better seen.

4. CONVOLUTIONAL NEURAL NETWORK (CNN)

Convolutional neural networks are a class of networks for deep learning specialized for working with images. They consist from input layer, hidden layers and output layer; layers which responsible for convolution are inside the hidden layers. By convolution of input matrix and kernel, we get a new matrix that has more pronounced aspects useful for further learning, which is defined as

$$(A * B)_{i,j} = \sum_{k=0}^{p-1} \sum_{l=0}^{q-1} A_{i+k,j+l} B_{k,l} \; i = 0, \dots, m-p; j = 0, \dots, n-q$$

where is $A \in R^{m \times n}$ input matrix (images or another signal) and $B \in R^{p \times q}$ is kernel (or filter).

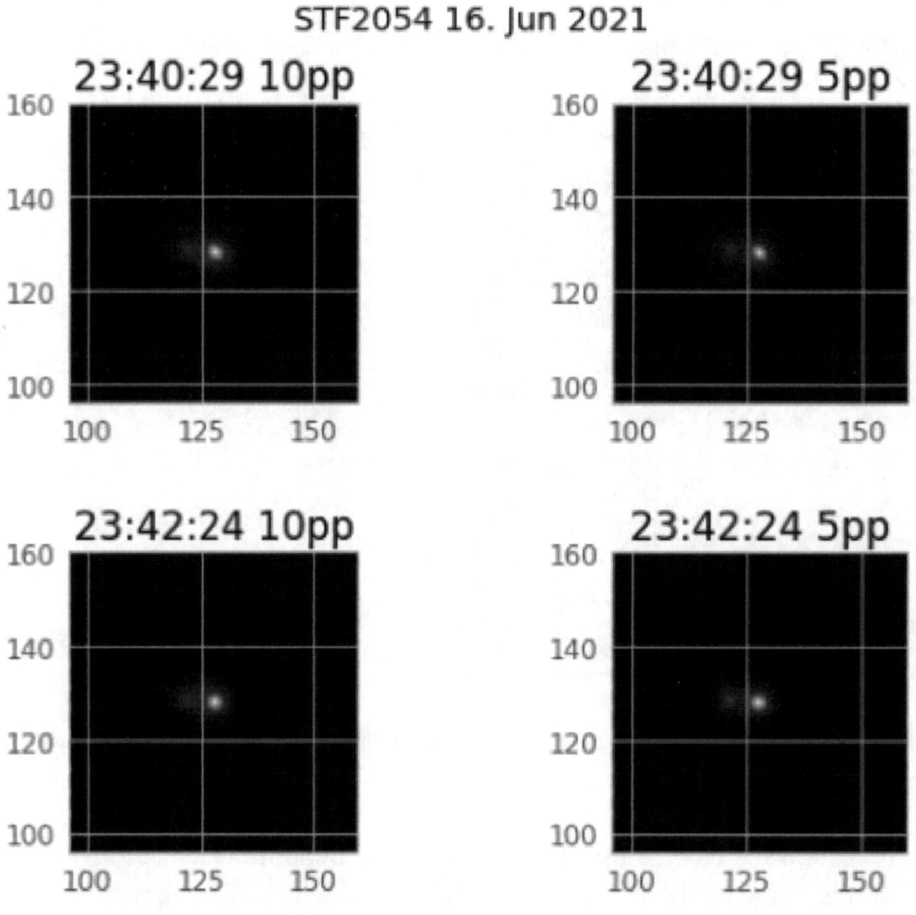

Figure 3: Four acquired frames of WDS 16238+6142 = STF 2054AB in reduced size (94 x 94 pixels).

A convolutional network is trained to minimize output errors of given inputs from the training set. During that process, kernels are learned that implement useful transformations over the input signal. In the general case, CNN works on tensors (arrays of matrices of the same dimensions), where a single matrix is a channel.

Accordingly, tensor convolution operation is defined in similar way as earlier but $A \in R^{m \times n \times c}$ and $B \in R^{p \times q \times c}$ are input tensors, and as output we get a matrix

$$(A * B)_{i,j} = \sum_{k=0}^{p-1} \sum_{l=0}^{q-1} \sum_{t=0}^{c-1} A_{i+k,j+l,t} B_{k,l,t} \; i = 0, \dots, m-p; j = 0, \dots, n-q.$$

CNN learns a certain number of filters in each layer. A set of filters that act on the same input signal are called convolutional layer; it outputs a tensor so to each channel corresponds output of one filter. The outputs are further non-linearly transformed using activation function, most often ReLU (Rectified Linear

Unit) but we used a slightly more common Swish (Ramachandran et al. 2017), by applying the given activation function to each tensor channel.

Another type of frequent layers used in CNNs are pooling layers. These layers perform aggregation data in order to reduce calculations and the number of network parameters. They are realized by changing part of tensor channel in certain dimensions with one value, in relation to the type of aggregation which we use. The most common type of aggregation is maximum pooling; with this, high values are preserved i.e. patterns detected by kernels, but is lost its specific position. The importance of this compromise i.e. model size/training time and accuracy depends on specific applications.

5. SHORT DESCRIPTION OF USED MODEL

The convolutional network was implemented in Python using Keras[2] and the Scikit-learn[3] library. The input data is simulated signals, described in section 2, in the form of tensors with dimension 256x256x1 (one-channel tensor). The output layer consists of only two neurons: one for angular separation and the other for position angle. Division of the set into training, test and validation subset was performed randomly, in the ratio 80:10:10, respectively.

Minmax normalization was applied to the recorded frames because of improving performance and stability of the model

$$x_{norm} = \frac{x - x_{min}}{x_{max} - x_{min}}$$

where x_{min} and x_{max} are minimal and maximal pixel intensity on the frame. In order to avoid overfitting and improving generalization, we used two regularization techniques: early stopping and the checkpoint model. When we train a model, an insufficient number of epochs leads to underfitting, but an excessive number leads to overfitting. Compromise is an early stop - an

[2] https://keras.io/

[3] https://scikit-learn.org/stable/

interruption of training at the moment when performance of validation set begins to decline. During training process, we monitor the validation loss and we remember those model wich have currently the smallest validation error. As the error function for our model, we used standard mean squared error (MSE).
The complete calculation process is shown by the block diagram in Fig. 4.

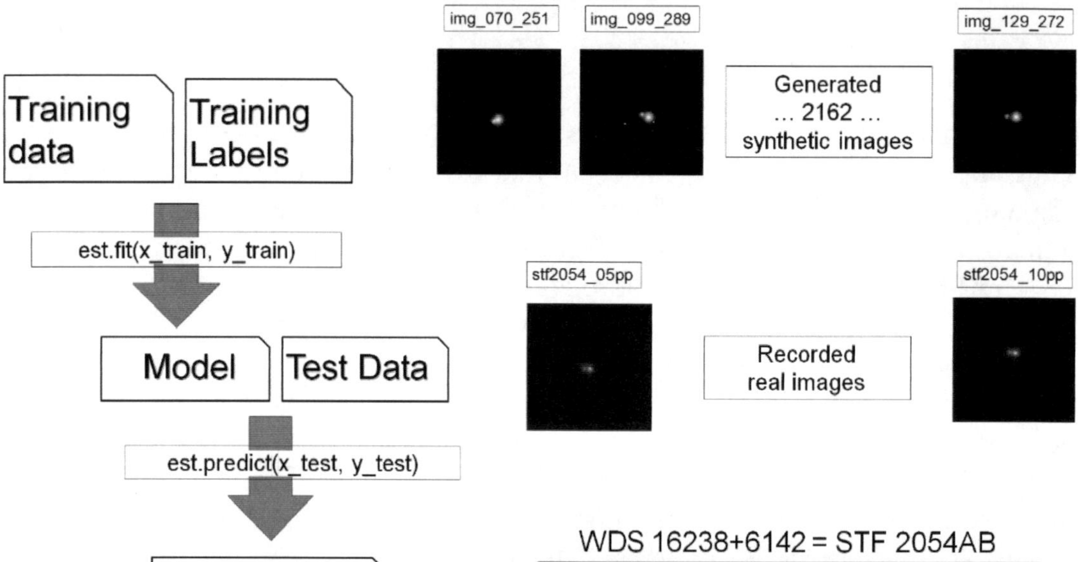

Time of observation	Number of the best frames	Angular separation (")	Positional angle (°)
23:40:29	5%	0.9821	259.35
	10%	0.9827	259.86
23:40:29	5%	0.9742	260.89
	10%	0.9773	260.53

Figure 4: Block diagram of the calculation process.

6. CONCLUSIONS

The calculated MSE for our model was $4.52 \cdot 10^{-6}$ which indicates stable and accurate model. The final values for the separation and position angle with the corresponding standard deviations for double star WDS 16238+6142 = STF 2054AB are

$$\rho = (0.9791 \pm 0.0040),\ \vartheta = (260.16 \pm 0.69)^{o}.$$

The obtained small values of the standard deviation for separation and positional angle indicate that convolutional neural network can be applied to "measure" these quantities.

References

Pavlović R., Cvetković Z., Damljanović G., Jovanović M.D.: 2021, Lucky Imaging at Vidojevica, *Publ. Astron. Obs. Belgrade*, 100, 339-343.

Ramachandran P., Zoph B., Le Quoc V.: 2017, *Searching for Activation Functions* https://arxiv.org/abs/1710.05941

Proceedings of the XIII Bulgarian-Serbian Astronomical Conference (XIII BSAC) Velingrad, Bulgaria, October 3-7, 2022
Editors: Evgeni Semkov, Milan S. Dimitrijević, Momchil Dechev and Zoran Simić
Publ. Astron. Soc. "Rudjer Bošković" No 25, 2023, 143-149

OPTICAL OBSERVATIONS OF SOLAR CHROMOSPHERE IN NATIONAL ASTRONOMICAL OBSERVATORY

NIKOLA PETROV, TSVETAN TSVETKOV and YOVELINA ZINKOVA

Institute of Astronomy and National Astronomical Observatory, Bulgarian Academy of Sciences
72 Tsarigradsko shose blvd., Sofia 1784, Bulgaria
E-mail: tstsvetkov@astro.bas.bg

Abstract. There is hardly a person nowadays who has not heard and is not concerned about the growing climate problems threatening the life of most plant and animal species on Earth, including the humankind. We understand more and more that one of the important factors determining the global trends of climate changes on Earth is the cosmic influence on the magnetosphere, atmosphere and biosphere of our planet. But of all the cosmic factors, the Sun and solar activity are undoubtedly the most important one and play a crucial role for all processes on Earth. This is the main reason for the widespread scientific interest in studying and understanding the physics of the Sun. New both ground- and space-based observing programs that aim to observe and study solar activity are being introduced every year. Exploration of solar chromosphere, where many active processes take place and new ones often emerge, is among the important problems that ground-based telescopes are capable of solving. It is also one of the main tasks of the of Sun and Space Weather Department at the Institute of Astronomy and National Astronomical Observatory, Bulgarian Academy of Sciences. The long-term efforts of our entire team led to the financing, construction and establishment of the first chromospheric telescope in the National Astronomical Observatory Rozhen. We present the main parameters and possibilities of our new 30-cm chromospheric telescope.

1. INTRODUCTION

Solar chromosphere ("color sphere") is situated directly above the visible surface of the Sun, called the photosphere ("light sphere"), which provides the light which is dominant for human vision. Except for the moments right before and after the totality during solar eclipses when the chromosphere could be seen as a thin purple-red ring surrounding the Moon's dark disk (Figure 1), it remains invisible to the human eye. Since it is located immediately above the point of temperature minimum in the solar atmosphere, the density in its lower part decreases rapidly with height. Therefore, the general reduction in density (10^{-4} times) is also large, which explains many of the characteristics of the chromosphere.

Figure 1: The solar corona and the chromosphere as photographed during the third contact of the total solar eclipse in 2017 (USA).

Studying the chromosphere, we are capable of finding answers to the questions connected to wave propagations (Mullan D. J., 2010). When high-frequency sound waves originating in the turbulent convective zone penetrate the chromosphere, they undergo two changes that are responsible for the heating of the chromosphere from 4300 K to almost coronal values:

- Due to the rapid drop in density, they accelerate and become shock waves. More energetic particle impacts lead to stronger heating as the wave passes.
- The reduced density of matter and the presence of magnetic field increases energy transfer by MHD waves. They propagate along the magnetic field, induce oscillations of the charged particles, which leads to more collisions between them and with neutral particles and finally the energy of the wave is converted into heat.

On the other hand, there are numerous active manifestations of solar activity that determine space weather and thus influencing the Earth's climate (Nikolov & Petrov, 2014; Pomoell & Poedts, 2018) – flares, coronal mass ejections, bulges/filaments, solar energetic particle events (Wang et al., 2003; Jiang et al., 2014, Tsvetkov et al., 2019).

2. 30-CM CHROMOSPHERIC TELESCOPE

A brand new Schmidt-Cassegrain telescope with 305 mm aperture and 3050 mm focal length was installed in the solar tower of National Astronomical Observatory (NAO) in 2022 (Figure 2).

Figure 2: 30-cm chromospheric telescope, installed in the solar tower in National Astronomical Observatory.

It is equipped with K8 flat glass infrared (IR) filter with bandwidth 6560±500 that reflects more than 94% of the electromagnetic emission in the non-working regions of the spectrum to prevent superheating of the telescope (Figure 3). To reach the desired and necessary selectivity (passband), thin layers of various metals and their oxides are applied successively on a glass pad. The IR filter is covered on both sides with a layer of quartz, which is also a protection against atmospheric microparticles or dust. Unfortunately, sensitivity to moisture is a common property of many metals and their compounds, but filters larger than 30 cm in diameter that are insensitive to moisture are still difficult to produce, which raises their price repeatedly.

Figure 3: The infrared filter of the 30-cm chromospheric telescope.

Before reaching the focal point behind the primary mirror, the light also passes through a system of telecentric lens extending the focal length, 0.75 – 0.4× focal reducer, and Hα filter with bandwidth of about 0.3 Å. The focal length correction system determines an effective focal length of the telescope varying between 5000 and 15000 mm and an observational field of view below 10×10 arc min. This differentiates our telescope from the most common Hα instruments used nowadays as it provides a detailed view of selected areas of the Sun instead of full-disk observations.

3. OBSERVATIONAL CONSIDERATIONS AND POSSIBILITIES

As with any optical system, with the 30-cm telescope in NAO it is better to observe the central parts of the solar disk, rather than near the solar limb since the contrast of surface detail is much higher in the central regions. The intensity of the various manifestations of solar activity near the limb often shows higher amplitude, which necessitates at least 16-bits receivers. Of course, in order to avoid non-uniform illumination (vignetting or "light fall-off"), it is mandatory to make flat fields during every observation. Also, it would help to clear the images from the dust particles on the surfaces of the various optical elements or non-uniformity of transmission of the narrow-band Hα filter.

Receiving the raw images, it is better to have a lower illumination of the frame than a maximum filling of the digital matrix. The shorter the exposure, the more likely it is to get clear/contrasted photos. With shorter exposures (but more than a second), the perfect images can be obtained at the best seeing by removing the effect of atmospheric scintillations. The raw image capture technique we use involves video recording for 5-10 seconds. This allows taking 500-1000 single frames followed by a procedure of stacking the best quality images, where we can reach the maximum of the signal-to-noise ratio, as well as the best resolution of our optical setup – about 0.3".

Observing the Sun near the limb in Hα line (6563 Å), we see that the chromosphere is not a homogeneous region, but it consists of many separate bright line structures (Figure 4). They rise in height from the photosphere up to several thousand kilometers. They are called spicules and appear as spike-like formations in Hα. When detected on the solar limb, the spicules appear bright due to the lack of background light to be absorbed (the spicules are like emitting structures), while on the solar disk they are dark because of the absorbed the light of H-alpha photons.

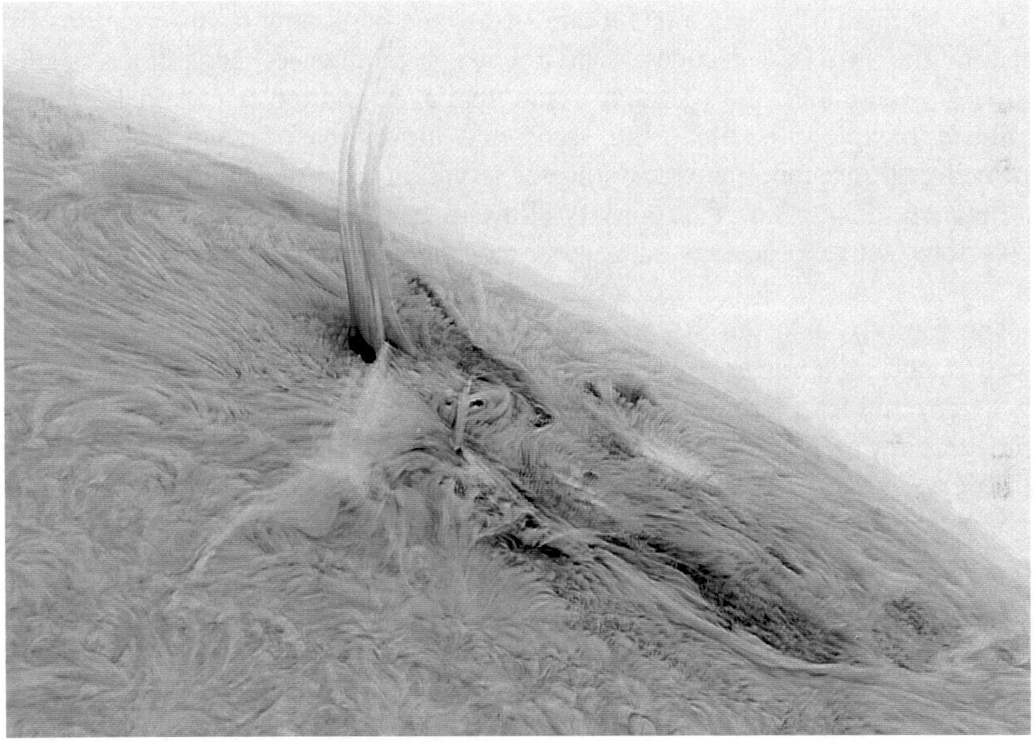

Figure 4: Image of an active region and chromospheric environment near the solar limb, obtained with the 30-cm telescope.

Another typical for the chromosphere formation could be noticed in Hα observations – chromospheric network. Their lifetime, appearance and transformation are still a matter of debate, which necessitates the constant monitoring of the active chromosphere on a large scale (full-disk solar observations). But it is also crucial to be able to obtain high-resolution observations so that the dynamics of the fine chromospheric structure can be determined more precisely. Of course, high-resolution observations are determined not only by the capabilities of optical instruments, but here the main role is played by the state of the atmosphere around the observers. Our observatory is located at altitude of 1740 m and offers good opportunities for regular observations as weather conditions in the observatory are assumed as appropriate for solar monitoring. These properties define our aim to determine the dynamics of the chromospheric fine structure with our new instrument. It may be the key to solving the physics of the interaction between the local magnetic fields of an active region and the solar plasma. Such interactions lead to the strongest manifestations of solar activity, such as prominence eruptions and CMEs.

4. CONCLUSIONS

Scientific solar observations passed through different stages over the years – visual observations, photographic methods, spectral detectors, coronagraphs, etc. The development of research techniques led to the beginning of observations from space in the last few decades, which have a number of advantages such as regularity, continuity, high temporal and spatial resolution. But still humankind continues to explore the Sun establishing new ground-based solar telescopes. The improvement of technologies for solar observations, the possibilities for constant and detailed monitoring of its activity allow us to further develop our ideas about the various active processes in solar atmosphere and the connections between them.

The recently installed 30-cm chromosperic telescope is the third instrument placed in the solar tower of Bulgarian National Astronomical Observatory after the 13-cm refractor from the beginning of the 1990s and the 15-cm solar coronagraph from 2005. Its scientific goals and objectives include offering regular Hα observations of the solar chromopshere as well as studying the evolution of large- and small-scale active structures like active regions, prominences/filaments, flares, spicules, etc.

Acknowledgements

This study is supported by the National Science Fund of Bulgaria with contract No. KP-06-H28/4 and the joint project of cooperation between IANAO, Bulgarian Academy of Sciences and the Department of Astronomy, Faculty of Mathematics, University of Belgrade, Serbian Academy of Sciences and Arts P-04/04.02.2020.

References

Mullan D. J.: 2010, *Physics of the Sun. CRC Series in Pure and Applied Physics*, ed. Dipak Basu, Taylor & Francis Group, LLC, pp. 219.

Jiang Y., Yang J., Wang H., Ji H., Liu Y., Li H., Li J.: 2014, Interaction and merging of two sinistral filaments, *Astrophys. J.,* 793(1), 14.

Nikolov T., Petrov N.: 2014, Main factors influencing climate change: A review, *Comptes Rendus de L'Academie Bulgare des Sciences*, 67(11), 1455-1476.

Pomoell J., Poedts S.: 2018, EUHFORIA: European heliospheric forecasting information asset, *J. Space Weather Space Clim.*, 8, A35.

Tsvetkov Tsvetan, Miteva Rositsa, Petrov Nikola: 2019, Filaments related to solar energetic particles, *AIP Conference Proceedings*, 2075(1), 090013.

Wang H., Qiu J., Jing J., Zhang H.: 2003, Study of ribbon separation of a flare associated with a Quiescent filament eruption, *Astrophys. J.*, 593(1), 564-570.

Proceedings of the XIII Bulgarian-Serbian Astronomical Conference (XIII BSAC) Velingrad, Bulgaria, October 3-7, 2022
Editors: Evgeni Semkov, Milan S. Dimitrijević, Momchil Dechev and Zoran Simić
Publ. Astron. Soc. "Rudjer Bošković" No 25, 2023, 151-156

BEAUTIFUL BUT NOT HOSTABLE – ORGANOLEPTIC PROPERTIES OF THE SOLAR SYSTEMS' PLANETS

BOYKO RANGUELOV

University of Mining and Geology, B. Kamenov str., Sofia 1700, Bulgaria
E-mail: branguelov@gmail.com

Abstract. The paper presents the summary of the works dealing with the organoleptic properties of the Solar systems' planets and some other space objects. The intensive development of the space research technologies, new and effective ways of communications and achievements of the space remote sensing methods, gives wide possibilities to investigate distant objects like solar systems' planets, moons, comets, etc. Obtaining data and information about the chemical composition, physical properties and other parameters of the space bodies and using methods to transform these properties in sound, taste, smell and other organoleptic peculiarities available only for the human senses, it is really interesting how the human beings will feel if they'll be able to explore the space bodies, by their own sensitive feelings. As a result of such investigation the conclusion that the planets of the Solar system looks sometimes very beautiful, but more often do not provide rather welcome environments for the human beings.

1. INTRODUCTION

The organoleptic properties of the space objects are subjective feelings depending of human perception. Some people like color, others – the dynamic of color changes, the subjective feeling of the human perception is different (for example https://www.google.com/search?channel=nus5&client=firefox-b-1-d&q=Pics+of+the+planets).

Some people are interested by other properties, like sound, smell, taste, fractality, etc. - see Ranguelov, Iliev (2019). On a first glance these subjective human feelings is difficult to determine quantitatively, but during the last years it is more and more an engineering problem – to transform the physical and chemical properties into human senses – Iliev, Ranguelov (2019).

The space exploration of the planets in the solar system and other astronomical objects is developing at a rapid and accelerating temporal domain. Missions with spacecraft orbiting closer and more distant objects are now a daily occurrence and bring more and more reliable information about various properties of the planets - see Ranguelov, Iliev (2019). One of the most interesting for the science are the taste, smell and possible sounds that can be felt during eventual visits to the various distant space objects - planets, satellites, comets, asteroids. Beautyfullness is a subjective feeling, and it is most often associated with the color range and dynamics of the layers enveloping the space bodies – Iliev, Ranguelov (2019).

At the basis of the methodologies for studying and collecting information which are well-developed and worked-out methods of physics, chemistry, astrophysics, remote measurements and other branches of science can help a lot to these achievements

The reliability of the obtained data is ensured by repeated missions of the spacecraft visits aimed at the study of the bodies of the Solar System and carried out by the various space countries. The interest of all developing space research countries in remote sensing and direct technologies is constantly growing and the number of missions is already over a hundred, which requires enormous scientific, technical and financial potential.

2. HOW THE INFORMATION IS COLLECTED?

The basis of these important studies is the chemical composition (elements and compounds) of the solid, liquid and gaseous envelopes of the studied space objects. Planetary chemistry methods are based both on remote methods from Earth and on the capabilities of the various analytical approaches used by spacecraft's (remote or direct sampling) as well as by direct astronauts' sampling (for example - on the Moon). These methods include several well developed and reliable techniques as spectral analysis (in transmitted or reflected light, in the infrared and/or ultraviolet spectral window), remote light and laser probing, gas chromatography, direct testing methods with various indicator reactions between chemicals sensitive to the action of one or another substance and other methods for determination of the chemical composition. The sensitivity of such methods is monstrous - concentrations of up to 1 ppm or even billionth parts of the substance can be determined (most people are familiar with this high sensitive analysis - when crossing the border points between countries, especially at the airports, the security people sometimes wipe sensitive tissue on clothes and/or other surfaces of our body and after seconds, the analyst says whether you have touched drugs - accidentally or on purpose). Similar high sensitivity methods are often used to determine the presence or absence of a substance (for example, in the atmosphere of planets, satellites or other space objects - comets, asteroids, etc.). The main way to collect information about the chemical composition (atoms, molecules and chemical compounds) and physical parameters (elevation, gravitational and magnetic field, radiation, etc.) are the space missions performed by various spacecraft reaching closer or farther of the studied space objects.

3. TRANSFORMATION OF THE CHEMICAL AND PHYSICAL PROPERTIES INTO ORGANOLEPTIC

After having a reliably determined chemical composition, the so-called transformation process, i.e. converting the data on the concentrations of various substances into an organoleptic sensation. This is especially true for detecting the smell of the space object. The cells in the human nose are specialized to sense the different odors of different chemical elements and compounds. The matter of taste is more complicated, but the mechanism of transformation works analogously. It is most difficult to transform the sounds emitted by the various cosmic objects into frequencies perceptible to the human ear. But the synthesizers of sounds easily cope with this task. After spectral analysis, shifting the frequency band to that perceived by the human ear is much easier. It should be noted that the lack of air (or atmosphere) around some space objects is not an obstacle to "sounding". Seismic waves, along with other elastic vibrations, are wave processes and can also be subject to spectral analysis and transformed placed within the audibility range of the human ear. Thus, all recordings of wave processes are transformed into sounds perceived by the hearing apparatus of people, creating the feeling of "sound of space objects". "Beautifulness" is dictated by the color spectrum, the mutual arrangement of colors on the images of objects, as well as of their change dynamics and variations.

4. ORGANOLEPTIC PROPERTIES OF THE OBJECTS OF THE SOLAR SYSTEM

The smell, taste, sounds and the "beautifulness" of the objects are extracted from their chemical composition and adapted for the human senses (illustrations – can be reached by a public domain located and visualized in the web-page https://www.google.com/search?channel=nus5&client=firefox-b-1-d&q=Pics+of+the+planets)

The Moon

The Moon is the only space body so far visited by space crews. Main chemical substances: Hydrogen, Helium, Oxygen, Sulphur, Iron, Silicates. Odor: Gunpowder (organoleptically detected by the astronauts. When the samples come into contact with oxygen – this smell disappears over time). Taste: matchsticks with a sour undertone. Sound: sharp tone resounding from the collected solid rock samples when kick on them, seismographs registered seismic waves (their "sound" is much longer than that one of the seismic sounds on Earth), the crackle sound as well as of meteorite impacts. Beautifulness: visible every evening from the Earth during the clear sky nights (even during the day). Color is gray brownish, elevation relief is full of craters, view, as of a volcanic desert. Weird lights probably generated by meteorite impacts.

Mercury

Chemical substances: Hydrogen, Helium, Sulphur, Silicates. Smell: probably like the Moon. Taste: sour. Sound: low-frequency echo (boom on impact). Beautifulness: color is brown-grey, elevation relief is full of craters, view is as a shining volcanic desert.

Venus

Elevation – heavily disintegrated with river valleys and plateaus; view - closer to Earth, predominantly desert, with mountains and plains, volcanic craters and meteorite impacts. Chemical substances: Sulfuric acid, Carbon monoxide and dioxide, Hydrogen sulfide, Phosphine, Silicates. Smell: Stifling, rotten eggs, rotten fish, highly poisonous atmosphere. Taste: intensively sour. Sound: of blowing winds and thunder, seismic waves observed. Beautifulness: color – silvery with brown components, thunder lights.

Mars

Elevation relief – craters, river valleys and the plateaus, strongly cut and with high amplitudes; view - closest to Earth, predominantly desert, with mountains and plains, volcanic craters and meteorite impacts. Chemical substances: Sulfur, Hydrogen sulfide, Iron and iron oxides, Silicates. Smell: Safety Matches, Gun powder, Rotten Eggs, Rust. Taste: of rust. Sound: of blowing winds, seismic waves, sometimes rolling stones. Beautyfullness: color - predominantly grey-reddish-brown.

Jupiter

Chemical substances: Hydrogen, Helium, Hydrogen sulfide, Ammonia, Phosphine, Hydrogen cyanide. Odor: Strong stench of rotten eggs, intestinal gases, urine, rotten fish, bitter almonds. Taste: sharp sour-bitter, poisoning. Sound: Roaring, low-frequency humming, frightening sounds. Beautifulness: color - silvery white and light brown and tan undulations from heavy storms and hurricanes, attractive.

Saturn

Chemical substances: Hydrogen, Helium, Hydrogen sulfide, Ammonia, Phosphine, Hydrogen cyanide, Ammonium hydrosulfide. Smell: Very strong stench of rotten eggs, intestinal gases, urine, rotten fish, bitter almonds, hair dye. Taste: sour-bitter, poisoning. Sound: Roaring, low-frequency humming, heavy wheezing. Beautifulness: color – bluish-brown, famous rings and many satellites (probably the most beautiful view of all extraterrestrial bodies).

Titan (The Saturn moon)

Chemical substances: Nitrogen (98%) Methylacetylene, Cyanoacetylene, atmosphere and liquid substances envelopes. Smell: Gas station. Taste: extremely unpleasant. Sound: whistling from geysers. Beautifulness: color - bluish, brown, violet, beautiful view, (probably most similar in colors to the Earth's).

Uranus

Chemical substances: Hydrogen, Helium, Hydrogen sulfide, Ammonia (less), Phosphine, Hydrogen cyanide, Ammonium hydrosulfide. Odor: Strong stench of rotten eggs, intestinal gas, urine, rot, bitter almonds, hair dye. Taste: sour-extremely unpleasant. Sound: No data (Saturn-like). Beautyfullness : color – bluish, view – attractive, with rings vertical on the rotation axis. Expectation of frozen sounds due to gravitational forces.

Neptune

Chemical substances: Hydrogen, Helium, Hydrogen sulfide, Ammonia (less), Phosphine, Hydrogen cyanide, Ammonium hydrosulfide. Odor: Strong stench of rotten eggs, intestinal gas, urine, rot, bitter almonds, marzipan, hair dye. Taste: sour-extremely unpleasant. Sound: No data (Saturn-like). Beautifulness: color – bluish, cold.

Are there any more unusual properties and effects in the Solar system?

They can be expected a lot. Of course, the study of the properties and characteristics of the bodies of the solar system cannot be limited to the present knowledge. There are a number of questions (a consequence of the current knowledge), which are yet to be solved (for example – the separate formation of solid and gas planets). The logic that denser chemical substances are building blocks for the planets closer to the Sun, and gaseous ones for the more distant ones, somehow does not coincide with the presence of Pluto and its largest moon, Charon, both solid bodies, the most distant from the Sun…or the biggest exception – the Earth, with its variety of chemical compositions, active geodynamics, hydrosphere and gaseous atmosphere. It is hardly to believe that was created by nature only to provide good conditions for the development of mankind. From what has been said so far, it follows that all the bodies of the solar system, with the exception of the Earth, are too unfavorable for humans and living organisms of the terrestrial type. So, LET'S SAVE OUR PLANET.

5. CONCLUSIONS

The organoleptic properties and beautifulness of the planets of the solar system is under discussion. Along this subjective feeling of the humans, some engineering problems arise to determine more important subjective properties like smell, sounds, even taste of space objects. Due to the engineering advances in probing space objects by spacecraft the convincing results are obtained due to the transformations of the chemical and some physical properties into the human feelings and senses. This branch of the knowledge seems to be in the frontiers of the near future space science. The analysis shows that the nice looking space bodies frequently are not very hostable to the humans. Even more – they are really dangerous for the potential visitors.

References

Ranguelov B., Iliev R.: 2019, Mercury's DEM and FAG fractal structure – indicator for meteorite bombardment by different density space bodies, *Russian Journal of Earth Sciences,* 19, 1-8.

Ranguelov B., Iliev R.: 2019, *Fractal Universe: A case study of Solar System.*, LAP Lambert Academic Publishing., Riga, 122 pp.

Iliev R., Ranguelov B.: 2019, Fractal properties of the Gas giants and their satellites within the Solar system, *To Physics Journal.,* 2, 3. 8-15.

http://purkh.com/index.php/tophy.

Proceedings of the XIII Bulgarian-Serbian Astronomical Conference (XIII BSAC) Velingrad, Bulgaria, October 3-7, 2022
Editors: Evgeni Semkov, Milan S. Dimitrijević, Momchil Dechev and Zoran Simić
Publ. Astron. Soc. "Rudjer Bošković" No 25, 2023, 157-163

THE OPTICAL PROPERTIES OF HYDROGEN PLASMA IN THE FRAME OF THE FULLY QUANTUM METHOD BASED ON A CUT-OFF COULOMB MODEL POTENTIAL IN DIPOLE APPROXIMATION

NENAD M. SAKAN[1], ZORAN SIMIĆ[2] and MOMCHIL DECHEV[3]

[1]*Institute of Physics, Belgrade University, Pregrevica 118, 11080 Zemun, Belgrade, Serbia*
[2]*Astronomical Observatory, Volgina 7, 11060 Belgrade, Serbia*
[3]*Institute of Astronomy and National Astronomical Observatory, Bulgarian Academy of Sciences, 72, Tsarigradsko chaussee Blvd. Sofia, Bulgaria*
E-mail: nsakan@ipb.ac.rs

Abstract. The absorption coefficients of hydrogen plasma, calculated within the frame of cut-off Coulomb potential model, for the wide area of electron densities and temperatures observed within the solar atmosphere are presented here. The optical parameter of hydrogen plasma of mid and moderately high nonideality parameter could be described successfully, thus enabling the modeling of optical properties, especially the calculation of plasma opacity. The model was proven in both convergence towards normal condition, ideal plasma case, as well as with the help of analysis of the experimental data and further theoretical consideration. The model potential is solvable in entire space and within entire energy spectrum, thus the yielded wave function solutions are a combination of a special functions. The special form of the cut-off Coulomb potential, possesses a unique feature that enables the precise, fully quantum method of calculation of inverse Brehmstrahlung effect. The method covers the plasma with nonideality coefficient Γ from 0.0005656 for $N_e = 2{\times}10^{18}$ cm^{-3} and T = 6 MK up to 2.69 for $N_e = 2{\times}10^{21}$ cm^{-3} and T = 10 kK, from non interacting plasma up to the systems with the strong Coulomb coupling. Although the presented method development is still a work in progress the possibility of unifying a mode for both transport and optical properties of plasma within same model is an attractive direction for its further development.

1. INTRODUCTION

In the theoretical models of Solar plasma a problems of plasma opacity, energy transport as well as radiative transfer under moderate and strong non-ideality are of strong interest. For the deeper analysis of the subject reffer to Fortov & Iakubov

(1999), Rogers & Iglesias (1998), Mihajlov et al. (2011, 2013). The strong coupling and density effects in plasma radiation were the subject of numerous experimental and theoretical studies in the last decades. The presented quantum way of describing atomic photo-absorption processes in dense strongly ionized hydrogen plasma is based on the approximation of the cut-off Coulomb potential. By now this approximation has been used in order to describe transport properties of dense plasma (see e.g. Fortov & Iakubov (1999), Mihajlov et al. (1989), Ignjatović et al (2017)), but it was clear that it could be applied to some absorption processes in non-ideal plasmas too Mihajlov et al. (2011), Mihajlov et al. (2011, 2015), Sakan et al. (2005). More detailed explanation could be find in Sakan et al. (2018).

2. THEORY

2.1. The approximation of the cut-off Coulomb potential

Many body processes could be described by the use of transformation to the corresponding single-particle processes in an adequately chosen model potential, For the detailed theory Sakan et al. (2018) should be considered. As an adequate model potential for hydrogen plasma of higher density, for instance reffer to Mihajlov et al. (1989), Mihajlov et al. (2011), the screening cut-off Coulomb potential, that satisfies above conditions, and is used here is in form

$$U_c(r) = \begin{cases} \dfrac{-e^2}{r} + \dfrac{1}{r_c} & : r \le r_c \\ 0 & r > r_c \end{cases} \tag{1}$$

where the mean potential energy of an electron in the considered hydrogen plasma $U_c = -e^2/r_c$ is used as an energy origin of the potential. Here e is the modulus of the electron charge, r - distance from the ion, and cut-off radius r_c - the characteristic screening length of the considered plasma.

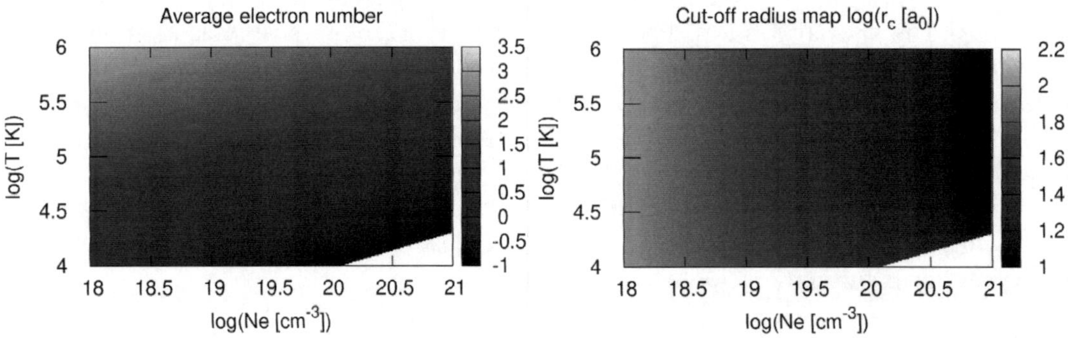

Figure 1: Left figure, the expected area of good agreement of the theoretical model, average electron number within screening sphere. Right, the expected area of good agreement of the theoretical model, logarithm of average electron number within screening sphere.

The cut-off radius r_c can be determined, see details in Mihajlov et al. (2009). The code that calculates the characteristic screening length of considered plasma uses electron density N_e, and temperature T, and is given at *https://github.com/nsakan972/ESPM-16.git*, it is free for use.

The model is expected to have good agreement with the modeled plasma conditions in the area where there is no many electrons within Wigner-Seitz sphere of screening radius r_c, and where the cut-off radius is smaller of few hundreds. The maps of the agreement could be seen on Figure **1**. In accordance with the presented data it is expected to have a good agreement of the model and theory within the range of electron densities $5{\times}10^{18}$ $cm^{-3} \geqq N_e \geqq 10^{20}$ cm^{-3}, and temperatures in range 10^4 K $\geqq T_e \geqq 10^6$ K. Such ranges could be related to Solar photosphere and part of the chromosphere, please refer to Fig. **1**. on page 347. in Massey (1982). Although there are evidences that the model describes the plasma of lower electron density numbers and lower temperatures, in such cases a more detailed models exist, and in our case a comparison that was carried out serves as a simple proof of validity of model.

The main quantyty that describes the non-ideality of plasma is a coefficient that presents a ratio of Coulomb energy to the termal energy, in accordance with Fortov & Iakubov (1999) is given by $\Gamma_{ee} = e^2\beta/r_{WS}$, where $\beta = 1/(k_B\ T)$, and r_{WS} is Wigner-Seitz radius. The values for Γ_{ee} are merrit for the Coulomb interaction importance, e.g. is it a close to ideal system or the plasma goes to strongly couplet coulomb system.

3. THE CALCULATED QUANTITIES

In accordance with previously mentioned theory the behavior of the dipole matrix element for the bound-bound transitions is investigated, and the results are presented here. The dipole moment is given by $D(r;r_c;n_i,l_i;n_f,l_f) = \langle n_f,l_f|\ \mathbf{r}\ |n_i,l_i\rangle$, where $|n_i,l_i\rangle$ and $|n_f,l_f\rangle$ are initial and final state wave functions obtained within the model of cut-off Coulomb potential. So all the photon emission and absorption parameters are related with the dipole matrix element. A essential calculation for this is the solving of the radial part of Schrodinger equation, e.g. finding of energy level values as well as the radial parts of the wave functions. The radial part, $R_{n,l;rc}(r)$ presented as $\chi(r) = r\ R(r)$ for the selected level n=3 and l = 0 as well as n=3 and l = 1 could be seen on Figure 2. left and right side figure respectively. Please note that in the strong local plasma field a wave functions differ significantly from Coulomb case and as such reflect on all parameters and further calculated values. The dipole moments for hydrogen atom without plasma influence are equal to the values from Hoang-Binh (2005). The behavior of dipole moments is investigated under the modeled plasma influence.

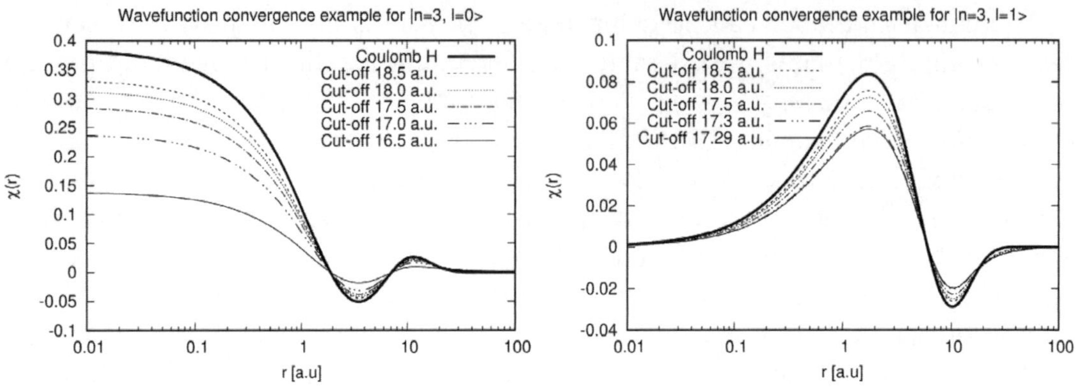

Figure 2: Left figure, the plasma influence onto the form of hydrogen wave function, n = 3, l = 0. Please note that $\chi(r) = r\,R(r)$, where R(r) is a radial part of wave function. Right, the plasma influence onto the form of hydrogen wave function, n = 3, l = 1.

The dipole moment is used for the calculation of all of the properties of dense plasma. For instance, as a illustration, the total absorption cross section is proportional to the dipole matrix element

$$\left(\sigma_0\right)_{\omega=\omega_{fi}} = \frac{1}{3}\frac{g_2}{g_1}\frac{\pi\omega_{fi}}{\epsilon_0 \hbar c}|D\left(r;r_c;n_i,l_i;n_f,l_f\right)|^2, \qquad (2)$$

and so any change of the wave function reflects on both dipole matrix element as well as total cross section symmetrically.

Table 1: The example set of dipole moments for hydrogen atom without plasma influence used for generation of behavior graphs on Figure 3.

Transition	$\lvert D\rvert^2$ $[a_0]$
|n=2,l=1> → |n=1,l=0>	1.29027
|n=3,l=1> → |n=2,l=0>	3.06482
|n=3,l=1> → |n=1,l=0>	0.516689
|n=3,l=0> → |n=2,l=1>	0.938404
|n=3,l=2> → |n=2,l=1>	4.74799
|n=9,l=1> → |n=1,l=0>	0.0820451
|n=9,l=1> → |n=2,l=0>	0.265475
|n=9,l=0> → |n=2,l=1>	0.0771136
|n=9,l=2> → |n=2,l=1>	0.314406

The test set of dipole matrix elements for the hydrogen atom in plasma calculated for the variety of cut-off radius, r_c, values is available at *https://github.com/nsakan972/ESPM-16.git*.

In order to have a correct modeling the influence of plasma microfield should be estimated. Within the presented model a cut-off radius r_c is used to model the plasma microfield. The investigation of the bound-bound transitions, i.e. the photoexcitation, is performed here.

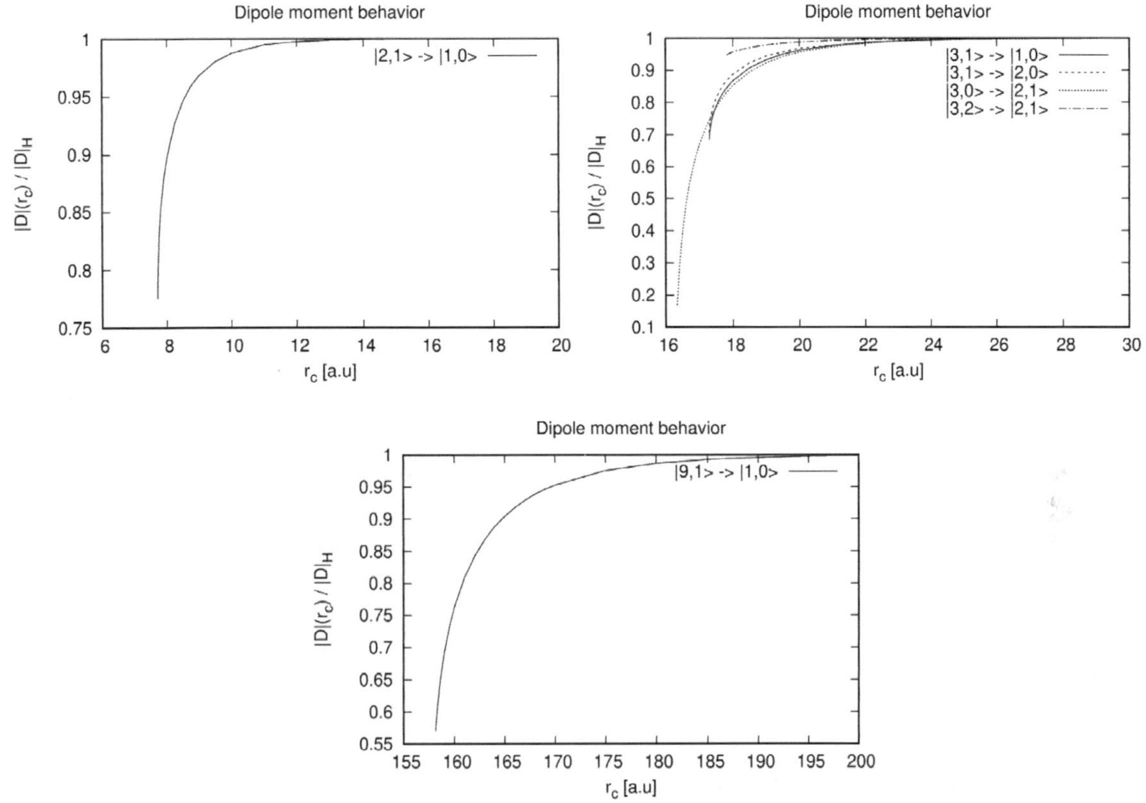

Figure 3: The behavior of normalized dipole moment regarding plasma influence for the transition |n=2,l=1> → |n=1,l=0>, left top graph. Plasma influence for the transition from upper levels with major quantum number n=3, right top graph. And plasma influence for the transition from principal quantum number n=9 to n=1, middle bottom graph.

As it could be seen from figures on Figure 3., the influence of the plasma micro field diminishes when the upper level is deep within the bound states energies, e.g. when it is burred deep within the bound states energies. When the upper level reaches the close vicinity of the boundary of the bound level states, e.g. if the realized bound state energy is close to zero in a potential given by equation (1), the plasma influence is no more neglectable.

It could be seen that the form of the dipole moment behavior is simple. It is a continuous and smooth function, so by knowing a cut-off radius at which the upper bound level starts to appear as well as a set of parameters for appropriate analytical function, a fast calculation of all related plasma parameters could be carried out. This gives an opportunity to perform a more complex modeling of a plasma screening boundary, at the vicinity of the model cut-off parameter r_c, and by such to ease the inclusion of more complex interactions, previously neglected by the model potential.

4. CONCLUSION

The presented work is a continuation of the previously developed modeling for the photoionization and inverse Bremsstrahlung processes of hydrogen plasma, and the goal is to include bound-bound processes within the same model. All of previously calculated data is also usable in modeling of Solar plasma processes.

In order to model a behavior of the plasma optical characteristics the used potential is a good approximation for the modeling of plasma interaction in a large area of densities and temperatures, details in Sakan et al. (2018). A strong plasma influence onto the optical parameters is observed where the plasma interaction energy is close to observed level energy value, e,g. when this level starts to appear, as illustrated in figures and analyzed within the manuscript. The work on using a more complex potentials is going on. We have tested a numerical method of wave function solution, and as a first step the Ar atom modeled is introduced, and initial values are in a expected range. As a second effect, a more detailed plasma-emitter interaction could be modeled. For both experimental praxis as well as Sun processes modeling an introduction of He atom and ion model is must, and as such could determine the further research.

Acknowledgments

This work is supported by Ministry of Education, Science and Technological Development, Republic of Serbia, Grant No. 451-03-68/2022-14/200024 (University of Belgrade, Institute of Physics, Belgrade-Zemun) and Grant No. 451-03-68/2022-14/200002 (Astronomical Observatory, Belgrade). One of the authors (MD) is thankful for the financial support from the Bulgarian Academy of Sciences (Bilateral grant agreement between BAS and SANU).

References

Fortov V., Iakubov I.: 1999, *The Physics of Non-ideal Plasma,* World Scientific https://books.google.rs/books?id=vFuZGoZsO2IC .

Hoang-Binh D.: 2005, A program to compute exact hydrogenic radial integrals, oscillator strengths, and einstein coefficients, for principal quantum numbers up to n ~ 1000, *Computer Physics Communications*, 166,(3), 191-196. https://doi.org/10.1016/j.cpc.2004.11.005

Ignjatović L., Srećković V., Dimitrijević M.: 2017, The Screening Characteristics of the Dense Astrophysical Plasmas: The Three-Component Systems, *Atoms*, 5(4), 42. https://doi.org/10.3390/atoms5040042.

Massey H.: 1982, *Applied Atomic Collision Physics, Atmospheric Physics and Chemistry*, Pure and applied physics, 43-1, Applied atomic collision physics, 1 edited by H. S. W. Massey, E. W. McDaniel, B. Bederson, Academic press.

Mihajlov A. A. Djordjević D., Meyer T. et al.: 1989, Determination of the electrical conductivity of a plasma on the basis of the Coulomb cut-off potential model, *Contributions to Plasma Physics,* 29(4-5), 441-446. https://doi.org/10.1002/ctpp.2150290414 .

Mihajlov A. A., Ignjatović L. M., Srećković V. A. et al.: 2013, The non-symmetric ion–atom radiative processes in the stellar atmospheres, *Monthly Notices of the Royal Astronomical Society*, 431(1), 589-599. https://doi.org/10.1093/mnras/stt187 .

Mihajlov A. A., Sakan N. M., Srećković V. A. et al.: 2011, Modeling of continuous absorption of electromagnetic radiation in dense partially ionized plasmas, *Journal of Physics A: Mathematical and Theoretical*, 44(9), 095502. https://doi.org/10.1088/1751-8113/44/9/095502 .

Mihajlov A. A., Sakan N. M., Srećković V. A. et al.: 2011, Modeling of the Continuous Absorption of Electromagnetic Radiation in Dense Hydrogen Plasma, *Baltic Astronomy,* 20, 604-608. https://doi.org/10.1515/astro-2017-0345 http://arxiv.org/abs/1112.6089 arXiv:1112.6089.

Mihajlov A. A., Srećković V. A., Sakan N. M.: 2015, Inverse Bremsstrahlung in Astrophysical Plasmas: The Absorption Coefficients and Gaunt Factors, *Journal of Astrophysics and Astronomy*, 36(4), 635-642. https://doi.org/10.1007/s12036-015-9350-0 .

Mihajlov A. A., Vitel Y., Ignjatović L. M.: 2009, The New Screening Characteristics of Strongly Non-ideal and Dusty Plasmas. Part 3: Properties and Applications. arXiv e-prints, p. arXiv:0906.5152. http://arxiv.org/abs/0906.5152 .

Rogers F. J., Iglesias C. A.: 1998, Opacity of stellar matter, *Space Science Reviews*, 85(1), 61-70. https://doi.org/10.1023/A:1005132518820 .

Sakan N., Srećković V., Simić Z. et al.: 2018, The Application of the Cut-Off Coulomb Model Potential for the Calculation of Bound-Bound State Transitions, *Atoms*, 6(1), 4. https://doi.org/10.3390/atoms6010004 .

Sakan N. M., Srećković V. A., Mihajlov A. A.: 2005, The application of the cut-off Coulomb potential for the calculation of a continuous spectra of dense hydrogen plasma, *Memorie della Societa Astronomica Italiana Supplementi*, 7, 221.

Proceedings of the XIII Bulgarian-Serbian Astronomical Conference
(XIII BSAC) Velingrad, Bulgaria, October 3-7, 2022
Editors: Evgeni Semkov, Milan S. Dimitrijević, Momchil Dechev and Zoran Simić
Publ. Astron. Soc. "Rudjer Bošković" No 25, 2023, 165-168

ACCRETION DISK UV LUMINOSITY OF QUASARS THROUGH SED FITTING

L. SLAVCHEVA-MIHOVA and B. MIHOV

Institute of Astronomy and NAO, Bulgarian Academy of Sciences,
72 Tsarigradsko Chaussee Blvd., 1784 Sofia, Bulgaria
E-mail: lslav@astro.bas.bg, bmihov@astro.bas.bg

Abstract. Large-scale structure at high redshifts is well traced by quasars. Among the most reliable distance indicators for them is the UV vs. X-ray luminosity relation. We aim at refining this relation by precise estimation of the accretion disk emission in the UV through decomposition of the spectral energy distribution accounting both for the quasar and the host galaxy. The three basic contaminators to the accretion disk luminosity in the UV are absorption, host galaxy emission, and broad-line region emission lines. The first two are accounted for by the spectral energy distribution fitting and the third by the usage of the numerous narrow-band filters of J-PAS. We present preliminary results for a couple of objects with data in J-PAS. The above approach for the UV vs. X-ray relation construction is expected to lead to reduction in the dispersion in the Hubble-Lemaître diagram.

1. INTRODUCTION

Quasars are robust tracers of large-scale structure up to high redshifts ($z \sim 7.5$). Cosmic voids and superclusters themselves can be used as probes of the dark energy component in the Λ-Cold Dark Matter (ΛCDM) model. The long-standing issue of the Hubble tension – over 4σ discrepancy between the estimates based on cosmic microwave background (CMB) under the assumption of the ΛCDM model, on one hand, and the ones involving the cosmic distance ladder method using Cepheids and supernovae, on the other (Verde et al. 2019) – remains unraveled. Variability is the main obstacle against using quasars as standard candles. The correlation between the luminosity in the UV and X-ray, reflecting the accretion disc vs. corona interplay, deserves attention due to its non-linearity and negligible redshift dependence. These can make possible the extension of the Hubble-Lemaître diagram (HLD) of supernovae to higher redshift ranges (Lusso et al. 2020; L20). For historical reasons, the correlation links the 2500 Å and 2 keV monochromatic luminosities. The above authors applied several filtering steps to produce a final cleaned sample with a drop in the relation dispersion. Its further reduction would be of high relevance for cosmology studies, in particular concerning the issues above.

2. AIMS AND APPROACH

The purpose of this study is to build a tight UV vs. X-ray relation for a quasar sample by the precise estimation of the accretion disk emission in the UV. It is most highly influenced by absorption and the contribution from the host galaxy and emission lines. We focus on the accurate estimation of the flux at 2500 Å through decomposing the spectral energy distribution (SED).

We compiled a subsample of L20's cleaned quasar sample with available data in J-PLUS, together with the few objects already present in MiniJ-PAS, and an overall dense SED coverage allowing the proper SED decomposition. The fitting was performed with the AGNfitter code accounting for the emission of the accretion disk, torus, host galaxy stellar population (SP), and cold dust related to the star formation regions (Calistro Rivera et al. 2016).

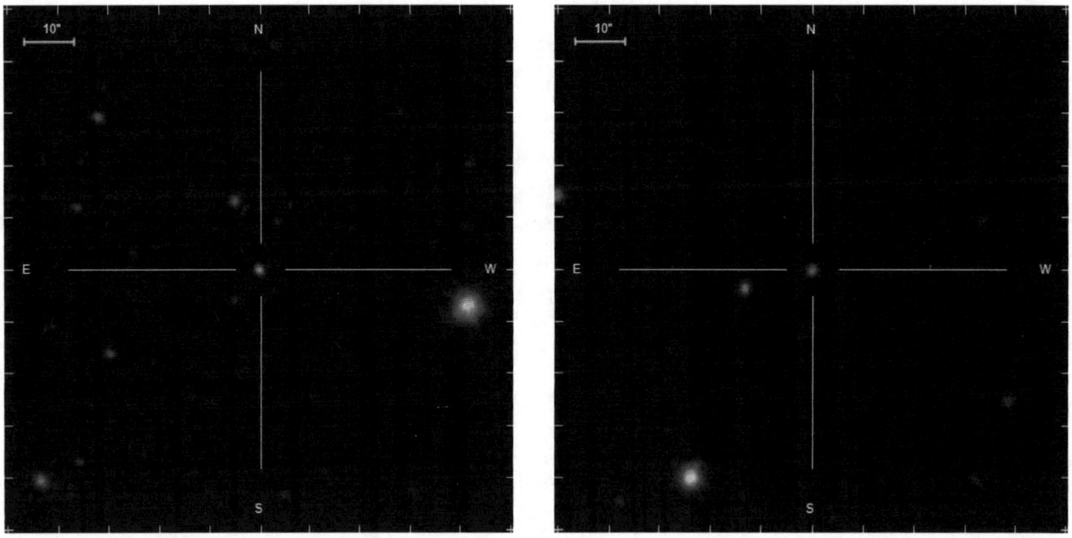

Figure 1: SDSS images of SDSS J000322.47-020051.6 (left) and WISEA J141816.23+522940.6 (right).

3. RESULTS AND DISCUSSION

As already noted, accretion disk emission in the UV is most highly influenced by three factors: absorption, host galaxy, and broad emission lines. The first two are accounted for by the SED decomposition. The usage of the numerous narrow-band filters of J-PAS enables the emission line flux specification.

We present preliminary results for a couple of objects with SDSS spectroscopic redshift estimates. The SED of the quasar SDSS J000322.47-

020051.6 (hereafter Q1; z = 2.03557 ± 0.00055) was built using data from WISE and J-PLUS. Data from WISE, MiniJ-PAS, and GALEX were used for the SED of WISEA J141816.23+522940.6 (hereafter Q2; z = 1.60524 ± 0.00021). Their SDSS images are shown in Fig. 1.

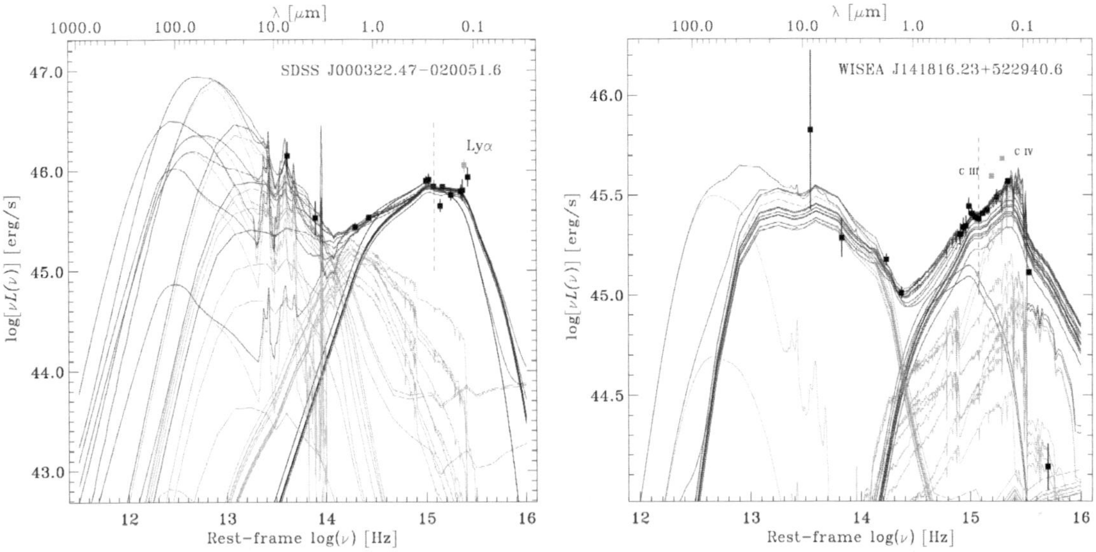

Figure 2: SED fitting with models accounting for the emission of the accretion disk (blue), torus (violet), host galaxy SP (orange), and cold dust (green). The green points correspond to the broad emission line flux mostly and were not considered in the fit. The dashed lines denote λ = 2500 Å. Left panel: SDSS J000322.47-020051.6. Right panel: WISEA J141816.23+522940.6.

Figure 2 shows the SED fitting for the two quasars. The flux data points corresponding to high contribution from the broad-line region emission (denoted) are excluded from the fit. The corresponding accretion disk flux at 2500 Å is $F_{AD, 2500Å} = 1.7 \times 10^{-28}$ erg $sec^{-1}cm^{-2}Å^{-1}$ and $F_{AD, 2500Å} = 8.7 \times 10^{-29}$ erg $sec^{-1}cm^{-2}Å^{-1}$ for Q1 and Q2, respectively. As can be seen from the decomposition, the bulk of the Q1 emission at 2500 Å is due to the accretion disk. Concerning Q2, the flux correction there is not negligible.

Building the UV vs. X-ray relation for a large number of quasars using accretion disk luminosities in the UV with the above corrections would result in a reduced dispersion in the HLD.

Acknowledgements

Funding for the SDSS and SDSS-II has been provided by the Alfred P. Sloan Foundation, the Participating Institutions, the National Science Foundation, the U.S. Department of Energy, the National Aeronautics and Space Administration, the Japanese Monbukagakusho, the Max Planck Society, and the Higher Education Funding Council for England. The SDSS Web Site is http://www.sdss.org/. This publication makes use of data products from the Wide-field Infrared Survey Explorer, which is a joint project of the University of California, Los Angeles, and the Jet Propulsion Laboratory/California Institute of Technology, funded by the National Aeronautics and Space Administration. GALEX is a NASA mission managed by the Jet Propulsion Laboratory. The main agencies supporting J-PAS are the Gobierno de Aragón, the Spanish MINECO, the Ministry of Science Technology and Innovation (MCTI) of Brazil, The National Observatory/MCTI, FAPERJ, FAPESP and CNPq.

References

Calistro Rivera G. et al.: 2016, *ApJ*, 833, 98.
Lusso E. et al.: 2020, *A&A*, 642, A150 (L20).
Verde L., Treu T., Riess A. G.: 2019, *Nature Astronomy*, 3, 891.

**Proceedings of the XIII Bulgarian-Serbian Astronomical Conference (XIII BSAC) Velingrad, Bulgaria, October 3-7, 2022
Editors: Evgeni Semkov, Milan S. Dimitrijević, Momchil Dechev and Zoran Simić
Publ. Astron. Soc. "Rudjer Bošković" No 25, 2023, 169-177**

STATUS OF VIDOJEVICA 20 YEARS AFTER ITS FOUNDING

OLIVER VINCE and MILJANA D. JOVANOVIĆ

Astronomical Observatory, Volgina 7, 11060 Belgrade, Serbia
E-mail: ovince@aob.rs, miljana@aob.rs

Abstract. Astronomical station Vidojevica was founded almost 20 years ago, i.e. in 2003. In this paper, we will present the twenty-year history of our station, its current status and future plans. Knowing about the plan to purchase a 1.5m telescope for the Rozhen observatory, we will especially refer to our 1.4m Milanković telescope, which was purchased from the same company - Astro Systeme Austria. Current observation projects and how to apply for the use of our telescopes will be also described.

1. INTRODUCTION

The Astronomical observatory in Belgrade (AOB), on the Zvezdara hill, had several excellent optical telescopes for observation in Yugoslavia at the time. Unfortunately, due to increasing light pollution, observations have become almost meaningless on most of our telescopes in the late 1980s. Only the high resolution Solar spectrograph was in regular use, the others being used only for educational and/or recreational observations. This led the management of the observatory to plan the construction of a new observation station. Around that time, extensive testing of various sites on observation conditions (seeing, sky transparency, etc.) began.

Due to historical circumstances, the implementation of the plan for the construction of a new station was full of challenges, so the location for the new observatory was defined and formalized only in 2003, when the property for the construction of the observation station was obtained from the state. Today, the station has three telescopes and several good instruments that serve to plan observations and monitor weather and observation conditions.

In the next section, we will describe a brief history from 2003. to the present day. Section 2. will describe our three telescopes with the instruments currently installed on them. Chapter 3. will describe how our observatory operate, as well as the way of allocating observation time on our telescopes. The last section will give a brief overview of future plans at the observatory.

2. SHORT HISTORY OF THE ASTRONOMICAL STATION VIDOJEVICA

As we mentioned earlier, the Astronomical Station Vidojevica (ASV) was founded in 2003. It is located on the Vidojevica mountain, which, geographically speaking, is located in the south of Serbia near the town of Prokuplje. Originally, the station had a dirt road and was connected to electricity, so the construction of elementary infrastructure proceeded relatively quickly. Already in 2005, a residence building was built, a well that provides drinking water was dug, and a pavilion for a 60 cm telescope was built. The further twenty-year history of the station can be presented in the following way:

- **2005, acquisition of the first telescope**. The first telescope that was acquired was a 60cm telescope from the Astro Systeme Austria (ASA) company. By the time of delivery, in 2005, we had already acquired a pair of modern CCD cameras and a portable spectrograph.

- **2010, installation of a 60cm telescope**. Due to various obstacles, the telescope was installed and calibrated only in 2010, and the following year we started with regular observations. Of the instruments used for monitoring the weather conditions at the station (and planning observations), we acquired and installed several instruments: DavisPro meteorological station, SBIG all-sky monitoring camera, and SBIG seeing monitoring camera. Of these instruments, only the seeing monitoring camera is not in use today, but several new instruments have been acquired in the meantime: Unihedron for measuring the brightness of the sky and AAG cloud-watcher for monitoring the cloudiness of the sky.

- **2010, the BELISSIMA project**. In the desire to acquire a larger and more modern telescope, a group of scientists at AOB submitted an application to the FP7 REGPOT call of the European Commission. The project called BELISSIMA (BELgrade Initiative for Space Science, Instrumentation and Modeling in Astrophysics) was approved and officially started on July 1, 2010. BELISSIMA[1] had 5 working packages, and one, the most important, worked on the procurement of 1.5m class telescopes (e.g. Samurović and BELISSIMA collaboration (2012), Samurović (2013), Samurović (2017)).

- **2016, installation of a 1.4m telescope**. As a final result of the BELISSIMA project, the telescope was procured from the ASA company, and it was delivered in May 2016. The telescope is installed in a pavilion with a sliding roof that was built for the temporary accommodation of the telescope. Thanks to the clear nights after the installation, the telescope was soon calibrated and put into operation.

---

[1] http://belissima.aob.rs/

- **2018, moving the 1.4m telescope to its pavilion**. Shortly after the installation of the 1.4m telescope, we started planning the construction of pavilion for the 1.4m telescope, the construction of which was completed in the middle of 2018 by a company from Serbia, while the dome was purchased from the Italian company Gambato[2]. After transferring the telescope to the pavilion, we soon started calibrating the telescope in order to again put the telescope into operation for astronomical observations as quickly as possible. This year was also marked by the formation of the commission for time allocation on our two telescopes, an obligation that was defined in the BELISSIMA project. In the same year, the ASV 60 cm telescope was moved to a pavilion with a sliding roof.

- **2020, installation of the 40cm telescope.** The Meade 40cm telescope was acquired back in 2004, and was in use at the AOB. It was transferred to ASV in 2019 with the intention of engaging the telescope into regular observations. The building of the pavilion was made by a company from Serbia, the same one that was engaged for the pavilion of the 1.4m telescope. The dome, on the other hand, was purchased from the Polish company ScopeDome[3].

- **2021, 60cm and 40cm telescope modifications**. A test of the 40cm telescope showed poor tracking, so Mead's original mount was replaced with a fork mount from Gemini Telescope Design[4] company in 2021. This year was also marked by the replacement of the old 60cm telescope mounting with the DDM200 from the ASA company.

- **2022, automation of the pavilion with a sliding roof**. After the decision to keep the 60cm telescope in a temporary pavilion with a sliding roof, the automation of the roof was done, so that it can now be controlled remotely.

As of the writing of this publication, ASV has three telescopes. The first one is the 1.4m telescope, which we named telescope "Milanković", after the famous

Serbian scientist Milutin Milankovic, who at one time was the director of AOB. The second telescope is a 60 cm telescope, which we named the "Nedeljković" telescope, also after the scientist who initiated the establishment of the observatory in Belgrade and who was chosen as its first director (e.g. https://www.aob.rs/Main/history.html). The third telescope is the 40cm telescope, which we unofficially call "György", after György Pál, who was the designer of the MoFoD assembly for our 40cm telescope[5]. While the 1.4m and 60cm

---

2 https://www.gambato.com/

3 https://www.scopedome.com

4 http://www.geminitelescope.com/

5 György Pál, of Hungarian origin, passed away in 2022. He personally installed the MoFoD mounting for our 40cm telescope at ASV, hence the initiative to call the 40cm telescope after him.

telescopes are in regular observational activity, the 40cm telescope is still in the test phase. In the following chapters, the main technical characteristics of all three telescopes will be described. Main characteristics of the site (seeing, weather conditions and so on) are described in the detail by Jovanović et al. (2012).

3. TECHNICAL CHARACTERISTICS OF THE TELESCOPES

3.1. The 1.4m Milanković telescope

The Milanković telescope is an Alt-Az mount with a Ritchey–Chrétien optical system. The diameter of the primary and secondary mirrors are about 1400mm and 460mm, respectively. Both mirrors are manufactured and tested at the LOMO company in Russia. Tertiary flat mirror enables directing the light beam to four outputs ports, two of which are Nasmith ports with a image-derotator that makes correction for image rotation in the focal plane of the ports. Effective focal distance (without additional optics) on all ports is about 11200mm. The optical system is designed to provide a 30 arcmin field of view without significant aberrations.

The telescope is powered by direct drive motors, which has numerous advantages over stepper motors. The number of parts that wear out is minimal with these engines, so their life is extended. The speed of starting and stopping the telescope is fast as well as its slewing to the celestial object (up to 6 degrees/sec). In addition, due to the small number of moving parts, the telescope is quiet when moving (admittedly, the telescope motors produces some resonant sounds that we could not remove when calibrating the telescope). Each of the motors, including the motors for the tertiary mirror, focusing with the secondary mirror and derotators, are equipped with high-quality absolute encoders. Altitude and azimuth encoders enable precise tracking (< 0.5 arcseconds in 10 minutes; tracking errors increase near the zenith, which is normal for this type of telescope mountings) and pointing (< 5 arcseconds of r.m.s pointing error).

The complete control system of the telescope is located inside the fork of the telescope and is controlled directly via the Internet. Currently, the computer is running under the Windows operating system. There is a wide variety of software that can be used to control the telescope and detector, but we have defined this selection as follows:

- **Autoslew**, is a program that is supplied by the ASA company and we use it to control the telescope. **Seqence** is another software from ASA that has various functions. Certainly the most important function of this software is the automatic creation of a pointing model, without which it would not be possible to precisely point the telescope at the given object. During regular observation we mainly use it for automatic telescope focusing, and sometimes for telescope guiding.

- **MaxImDL**[6] is the software we use to control the CCD camera and the filter wheel. It can also be used to control the telescope, and has other convenient possibilities, such as the transformation of pixel coordinates to WSC (using the PinPoint program), accurate centering of the target to the image center and so on. It allows connection to the DavisPro2 weather station, and records many meteorological parameters in the header of the FITS images.

- **ControlPanel** is a software developed by our collaborators, and it enables the control over the dome and its synchronization with the telescope. In addition to these most important functions, it also enables fan control in the pavilion, control over the humidifier, and basically control over all devices inside the pavillion remotely.

In principle, the above mentioned software packages are sufficient to run observations that are currently performed with our telescopes. However, **TheSky**[7] software is also used. It offers various functions, but we use it to take ephemeris of celestial objects in order to script observations. MaxImDL and TheSky provide an ActiveX interface that enables external control of telescopes, CCD cameras and other devices using scripting. ActiveX can be used with many programming languages (VBScript, JScript, Java, Perl, Visual Basic, Visual C++ etc.), but, in the Windows operating system, VBScript and JScript are most often used, because they can be run by the Windows Scripting Host, which is included in Windows.

As we mentioned, both Nasmith ports are equipped with an image derotator, and are used for observations (Figure 1.). Currently, one of the ports is equipped with an iKonL CCD camera and filter-wheel from the Andor company[8]. The 9-position filter-wheel is equipped with B,V,R,I standard Johnson-Cousin filters, L broad-band filter, and Ha, Ha continuum, SII narrow filters (~10nm) and are mostly used for photometric measurements and imaging.

The second port is currently equipped with two cameras and is specialized for lucky imaging of close binary stars (although, it is also used for other purposes such as tracking planetary occultation of distant stars). One camera is the iXon897 EMCCD[9] from the Andor company, which is characterized by fast recording and low readout noise. A x2 Barlow lens was placed in front of the camera to decrease the pixel scale for the Lucky imaging method. As a result, the field of view of this camera, with a pixel resolution of 512x512 pixels, is very small (~1.25 arcmin). Therefore, another CCD camera, SBIG STXL-6303[10], with a field of view of about 8.5x5.7 arcmin is installed on this port. The two cameras are used alternate-

[6] https://diffractionlimited.com/product/maxim-dl/
[7] https://www.bisque.com/product/theskyx-pro/
[8] https://andor.oxinst.com/products/ccd-cameras
[9] https://andor.oxinst.com/products/ixon-emccd-cameras
[10] https://diffractionlimited.com/product/stxl-6303/

Figure 1: The 1.4m telescope (middle) and the two ports with image-derotator.

ly with the Perseus 4-Port Instrument Selector[11] from the Optec company, which may redirect the light beam to four different outputs.

At the beginning of the use of the telescope, a SpectraPro 2700 portable spectrograph was installed on this port. However, due to the large difference in the f-number of the telescope and the spectrograph, the spectrograph was only useful for the spectroscopy of bright objects, i.e., < 6-8mag (e.g. Vince et al. (2018).

Also, before the Lucky imaging instrument set, the telescope had an improvized Savart plate polarimeter installed for polarimetry testing. Except for educational purposes, this polarimeter has not entered regular use. Nevertheless, its use encouraged us to include the polarimeter in our future plans for this telescope.

3.2. The 60cm Nedeljković telescope

The 60cm telescope, later named Nedeljković, has been in operation since 2010 and there are numerous publications where details about this telescope can be found (e.g. Vince & Jurković (2012)). Here we will only briefly describe its features and current instrumental setup.

The 60 cm telescope was also purchased from the ASA company. It is an equatorial mount with a Cassegrain optical system (Figure 2.). The mirrors were also made in the LOMO factory and are of excellent quality. The theoretical focal distance of the optical system is 6000 mm. The telescope is supplied with a x0.5 focal reducer which doubles the field of view but distorts the PSF of the stars near the edge of the field of view, so it is not in a regular use. Currently, the telescope

[11] https://optecinc.com/astronomy/catalog/perseus/default.htm

is equipped with the FLI PL230 back-illuminated CCD camera[12] of Finger Lakes Instrumentation, and a filter-wheel with Johnson-Cousin B, V, R, I and SDSS u, g, r, i, z, y filters. As far as observations are concerned, photometric observations in the mentioned filters are mainly done on this telescope.

Figure 2: The 60cm telescope.

As mentioned, due to problems with the dome of the pavilion, the 60cm telescope was moved to the pavilion with a sliding roof. The telescope is optimally positioned, so that it is protected from the wind and has a good view of the sky. The pavilion has shown all its advantages (e.g. quick equalization of the temperature of the surrounding objects with the ambient temperature and stabilization of seeing) and disadvantages (e.g. rapid dewing of telescopes and instruments if the humidity in the air is relatively high), and currently there are no plans to restore the dome of the 60 cm telescope and return the telescope to the old pavilion. The pavilion has recently been automated, so remote control of the roof via the Internet is possible. Likewise, the old assembly was recently replaced with the DDM200 (made by ASA company), which is based on direct drive motor technology.

3.3. The 40cm MEADE telescope

Although the Meade 16" telescope was the first of the ASV telescopes to be acquired, it was the latest to be installed. In fact, due to the numerous problems we still encounter today, the telescope is still in the test phase. Here we will briefly describe its main features.

[12] https://www.flicamera.com/proline/

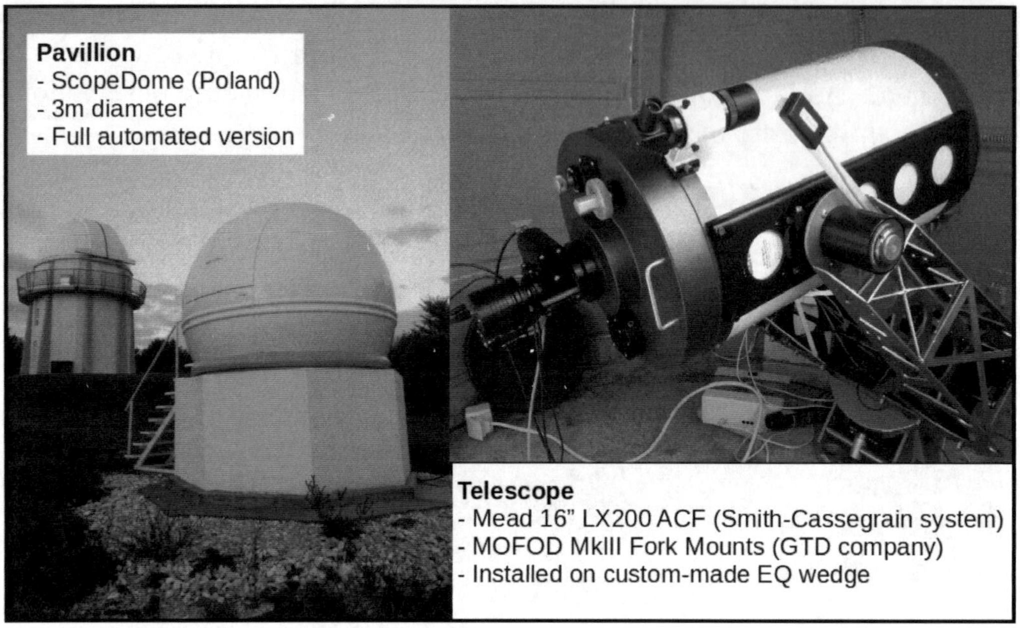

Figure 3: The 40 cm telescope.

The 40cm telescope is an older Meade product - specifically the Meade 16" LX200 ACF, with a relative aperture of f/10, i.e. a focal length of about 4000mm. The telescope is a Smith-Cassegrain optical system. It was delivered with a tripod with the aim of being used as an Alt-Az system, however, we placed the telescope on a concrete pillar and calibrated for using it as an equatorial fork system (mainly due to the intention of using it for long exposures without an image derotator).

The telescope's original mount showed very poor tracking which was probably the result of some damage to the motor and transmission system during the transport of the telescope. Therefore, a new mount was provided from the Gemini Telescope Design company- Gemini MOFOD MkIII Fork Mounts[13]. The telescope is currently equipped with the SBIG STXL-6303 camera for testing.

4. MODE OF OPERATION OF THE ASV

Currently, the number of people serving the ASV is relatively small - a manager, two technical operators and one janitor. Considering the current situation, observations are made only for 15 days around the new moon. Observations are made by astronomers who have been given time on the telescope and who receive a short training session before using the telescopes. Technical operators are responsible for the training of new observers, as well as the proper operation of telescopes and instruments that are used at a given moment.

Observation time is allocated based on applications, which are evaluated by the Time Allocation Committee (TAC). Applications are submitted twice a year.

---

[13] http://www.geminitelescope.com/mofod-mkii-fork-friction-mount/

On average, about 20 observing applications per semester are submitted to both telescopes. All details related to the submission of applications and the functioning of the TAC are described in the regulations that are publicly available on our official web address: http://vidojevica.aob.rs/.

5. CONCLUSIONS AND FUTURE PLANS

In the 20 years of existence of the Astronomical Station Vidojevica, it has become a productive observation station that has three telescopes fully equipped with detectors for photometric and astrometric observations. The observatory is also equipped with various instruments that are necessary for planning and conducting observations. As for the allocation of observation time and the observation itself, they follow the rules that spontaneously formed over time. Although these rules are defined in various regulations, they change over time (for example, with some major changes in the operation of the station).

As for the future plans, these are the most important ones for the near (distant) future:

- Testing of the 40 cm telescope and its introduction into regular observational work.
- A year ago, equipment was acquired for alternative supply of electricity to the station. The equipment consists of two wind generators and a network of solar panels that will be installed at the station next year. The basic idea with this equipment is to bridge long periods of time without electricity at the station.
- We have been planning instruments for spectroscopy and polarimetry for a long time. In cooperation with Russian colleagues from the Special Astrophysical Observatory, Russian Academy of Science, we created a conceptual project for an instrument that would be used as a low-resolution spectrograph and imaging-polarimeter.

References

Jovanović et al.: 2012, *Publ.Astron.Obs.Belgrade*, 91, 83.
Samurovic, S.; Belissima Collaboration: 2012, *Publ.Astron.Obs.Belgrade,* 91, 71
Samurovic, S.: 2013, *Publ.Astron.Obs.Belgrade*, 92, 11
Samurovic, S.: 2017, Publ.Astron.Obs.Belgrade, 96, 323
Vince, O., Jurković, M.: 2012, *Publ.Astron.Obs.Belgrade*, 91, 77.
Vince, O. et al.: 2018, *Publ.Astron.Obs.Belgrade*, 98, 341.

Proceedings of the XIII Bulgarian-Serbian Astronomical Conference
(XIII BSAC) Velingrad, Bulgaria, October 3-7, 2022
Editors: Evgeni Semkov, Milan S. Dimitrijević, Momchil Dechev and Zoran Simić
Publ. Astron. Soc. "Rudjer Bošković" No 25, 2023, 179-186

THE ACTIVE ZONE MOBILITY IN A MAGNETIZED DISK WITH ADVECTION

KRASIMIRA YANKOVA

Space Research and Technology Institute BAS, Sofia, Bulgaria
E-mail: f7@space.bas.bg

Abstract. In this paper, we consider some active processes in the accretion disk of a compact object. An active zone and its construction in various objects is a consequence of the advection action in the disk. We analyses the active zone's location and its development in the disk. We estimate an alteration in the active zone's behavior, throughout the disk's outer to the inner parts, for different astrophysical sources. We discuss the similarities and differences.

1. INTRODUCTION

In series of papers (Yankova, 2007-2015) we created a new magneto-hydrodynamical model of accretion discs, based on a new specific advective hypothesis, presented in Yankova (2013, 2015a).

$$\frac{\partial \rho}{\partial t} + \nabla.(\rho v) = 0 \qquad \nabla.v = 0 \qquad \nabla.B = 0$$

$$\frac{\partial v}{\partial t} + v.\nabla v = -\frac{1}{\rho}\nabla p - \nabla \Phi + (\frac{B}{4\pi\rho}.\nabla)B + \vartheta \nabla^2 v$$

$$\frac{\partial B}{\partial t} = \nabla \times (v \times B) + \eta \nabla^2 B \qquad \eta = \frac{\eta_m}{\rho} = \frac{c^2}{4\pi\sigma}$$

$$\rho T \frac{\partial S}{\partial t} - \frac{\dot{M}}{2\pi r} T \frac{\partial S}{\partial r} = Q^+ - Q^- + Q_{mag}$$

$$p = p_r + p_g + p_m$$

Here v is velocity of the flux; ρ - mass density; B - magnetic field; Φ- gravitational potential; p-pressure; - ϑ - kinematical viscosity; -η- magnetic viscosity; Q^+ - viscosity dissipation; Q_{adv}– advective term; Q^- - radiative cooling. The unified model we built on the conception of the non-deforming advection, allows in the field of nonlinear physics to conduct an analytical study of the accretion.

The researching of the emerging because of the advective mechanism, direct and reverse connections in the disk flow, represent a solution to the global model for the radial and vertical disk structures; the local structure model and adapted model for the emerging corona. Results in detail have given in (Iankova, 2007a, b), (Iankova, 2009), (Yankova, 2012a, b), (Yankova, 2015 b, c).

Here we use our work to analyze the state of the disk active zone and we will discuss the similarities and differences in both typical cases: a micro-quasar and a symbiotic recurrent nova.

In the papers (Yankova, 2015c, 2019) the disks structures of the high-mass X-ray binary system Cyg X-1 and the symbiotic binary star RS Ophiuchi have modelled. In the next sections, we will compare the active zone behavior: its width, separation, mobility and evolution in both objects disks.

2. ACTIVE ZONE

There are usually three zones of the activity in a disk: The first is in the outermost area and is associated with the inflow from a secondary source; The real active zone, where the activity resembles that in the inner layers of a star; And the activity in the innermost regions, which is practically coronary activity (by the disk's corona or the acretor's magnetosphere).

In this paper, our attention has especially focused on the active zone in an advective disk.

Advection have directly related to an appearance, development, behavior and movement of the active zone in the accretion disk.

As we have shown in our relativistic consideration (Yankova, 2017a, b), (Yankova, 2018a, b), the Advective Operator who consists by Pure Rotation+ Pure Translation in the space-time:

$$\left(\partial_{t_i} + \boldsymbol{v}_{ij}\partial_{x_j}\right)\boldsymbol{v}_{ji} = \beta_{ji}\partial_{t_i} + \delta_{ij}\partial_{t_j}, \tag{1}$$

Where

$$v_{ij} = \frac{\partial \chi_i}{\partial t_j}, \tag{2}$$

It naturally goes to the low-energy boundary of a non-deforming advection:

$$\frac{\partial(\rho \boldsymbol{v}_i)}{\partial \mathrm{t}} + \frac{\partial}{\partial \mathbf{x}_j}\left(\rho \boldsymbol{v}_i \boldsymbol{v}_j\right) = \rho\left(\frac{\partial \boldsymbol{v}_i}{\partial \mathrm{t}} + \boldsymbol{v}_j \frac{\partial \boldsymbol{v}_i}{\partial \mathbf{x}_j}\right) = \rho \frac{\boldsymbol{D}\boldsymbol{v}_i}{\boldsymbol{D}\mathrm{t}} \tag{3}$$

Earlier in (Yankova, 2015-2018) we have noted that incomplete modes of an advection (see radial advection Beloborodov (1998), Narayan&Yi (1994-95); or orbital advection see Fabian et al. 2012), have deformed the differential by an individual modification of one or another of its component. In this case, when the advection is a non-dominant mechanism, there are not any conditions of flow deformations. Then, a solution has transferred as a whole (eq.3).

The advection carries the properties of flow parameters from one area of the medium to another. The mechanism transfers (transports) all disturbances without deformations.

The active zone is determined by: - an outer radius, obtained from the condition $v_a^2 \leq v_s^2$, for the radius of the disk's corona (Iankova, 2007a). Where v_s and v_a are the sound and Alfven velocities in the disk flow. – An inner radius, obtained from the critical value of the heating K > 1, where K is a measure of the disc ability to qualitative the cooling (Iankova, 2009), or the condition $\langle v_a \rangle^2 \leq (9/4)\langle v_\varphi \rangle^2$ (Iankova, 2009), where v_φ is the angular velocity; and – the disk luminosity distribution.

The active zone have separated into two parts: the plateau in the luminosities (see Fig.3-4) indicates the outer active zone, in which the advection conceals the activity. In the inner active zone, the activity becomes observable (Yankova, 2019).

3. RESULTS

Our model could be applied to a large variety of the real objects, to explore various aspects their evolution. Here we obtain the active zones of the studied objects at two different evolutionary moments, separated by several periods $\sim\Omega_0^{-1}$:

For RS Oph the active zone is located in the ranges $\sim(0.008;0.5)R_0$, $R_{dstr}\sim(0.007;0.01)R_0$ (see Fig.1a) and $R_{out}\sim0.5R_0$ (see Fig.5). Here $x = r/R_0$ and $R_0 = 10^3R^*$ is the outer radius of the disc, measured in radii of the star- accretor;

For Cyg X-1 the active zone is located in the ranges $\sim(0.1;0.4)R_0$, $R_{dstr}\sim0.1R_0$ (see in Yankova, 2014) and R_{out} $\sim0.4R_0$ (see Fig.2b) respectively. Here $x = r/R_0$ and $R_0 = 10^3R_g$ is the outer radius of the disc, measured in Schwarzschild radii.

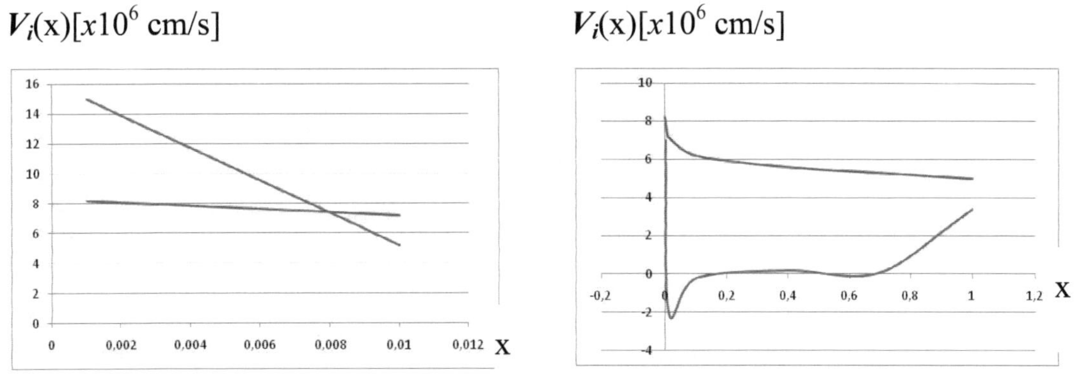

a) *b)*

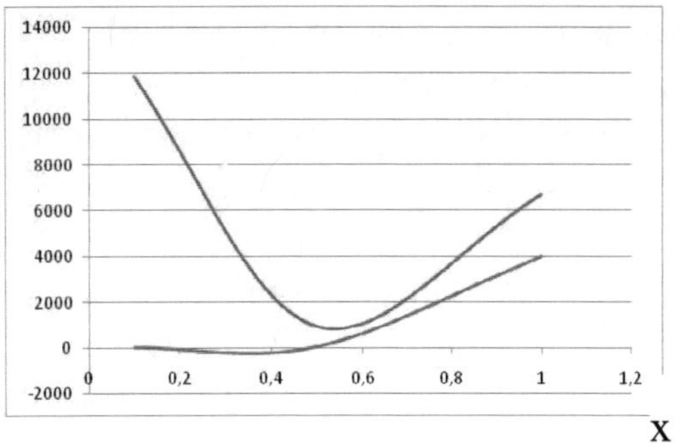

Figure 1: Distribution of the Condition $|v_a| \leq |(3/2)v_\varphi|$:
(a): for RS Oph, gives a destruction radius at ~8R*, and for Cyg X-1 (see Yankova, 2014) gives a inner radius at ~100Rg,
(b), (c): at the later stage, a destruction radius for both objects is not registered.

The active zone of RS Oph spreaa ınwards, in the ranges ~(0.005;0.5)R_0, R_{in} ~0.005R_0 (see Fig.1b, Fig.3) and R_{out}~ 0.5R_0 (see Fig.5);

While, for Cyg X-1 the active zone spread inward and outward, in the ranges ~(0.00?;0.9)R_0, R_{in} ~0.0?R_0 (see Fig.1c) and R_{out} ~0.9R_0 (see Fig.2c) respectively.

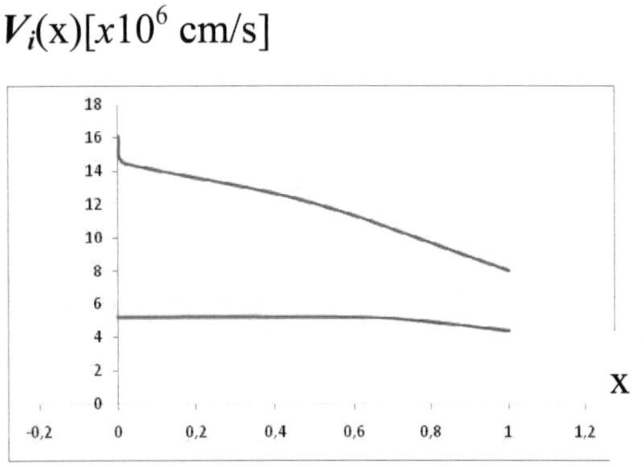

a1)

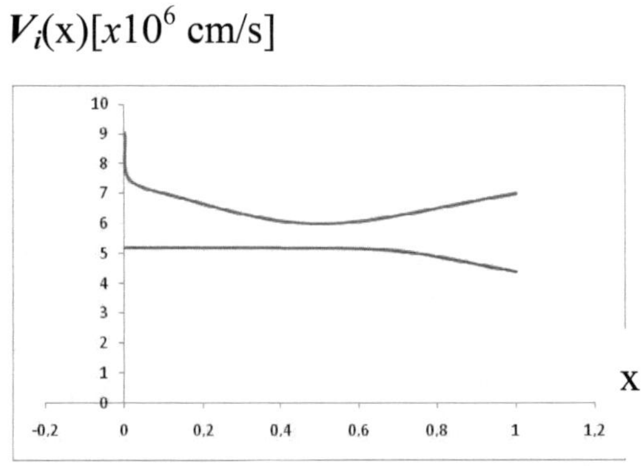

a2)

Figure 2(a): at RS Oph the disc does not generate a corona

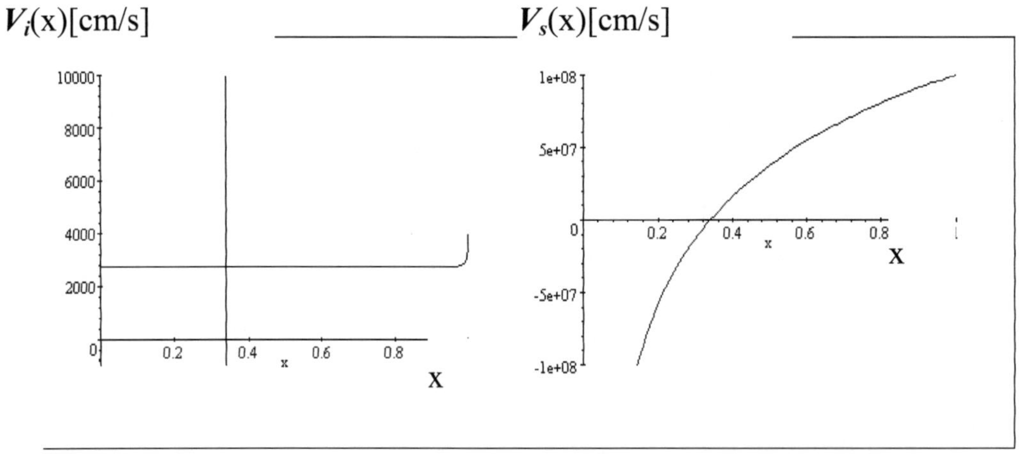

b)

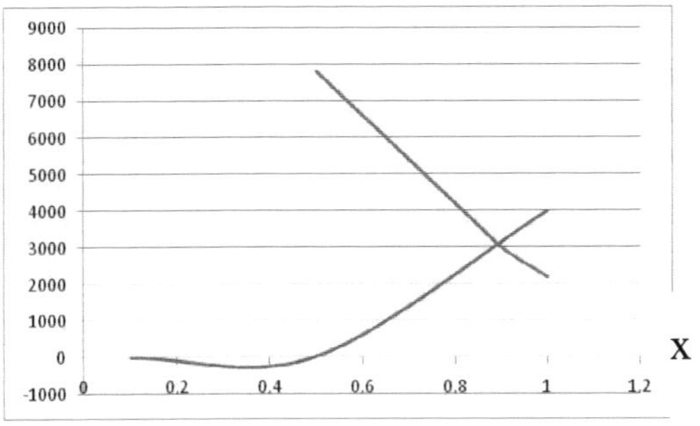

Figure 2: Distribution of the Condition $|v_a| \leq |v_s|$:
(b): for Cyg X-1 (Iankova, 2007) gives an outer radius at $\sim 340R_g$,and
©: at the later stage for Cyg X-1 gives an outer radius at $\sim 900R_g$.

4. ADVECTION IN THE DISK – – THE ACTIVE ZONE BEHAVIOR

The advection controls the flow parameters in both parts the active zone. In Newtonian flux, the advective ring in the disk is the main carrier of the hidden activity. In the advective ring, a physical effect of the local heating occurs, which represents the direct connection of the instabilities with the disk's energetics. The feedback is an expression of the hidden dependence of the internal flow structure of the nonlinear effects in it (Yankova, 2012a). The active zone is densely populated with rings in which the advection is reproduced. A self-induction mode is due to interaction of the electro-magnetic field with the plasma under the conditions of a strong gravity (Iankova, 2009; Yankova, 2015a).

Two factors determine the advection in the relativism: the magnetic field topology and gravity. The self-gravity in the formations, in which advection has generated; and background potentials, determined by space-time metrics are directly interrelated to the evolving advection (Yankova, 2018; Giuseppe, 2013). The magnetic fields include and determine the feedbacks of the mechanism of a fundamental advection. With its development in relativism, the advection works even better as a mechanism. Self-gravity literally closes the rings in the beds formed by the azimuthal current in the manifold topology.

The concentration of the rings determines the width and thickness of the active zone, and the advective displacement have related to its mobility.

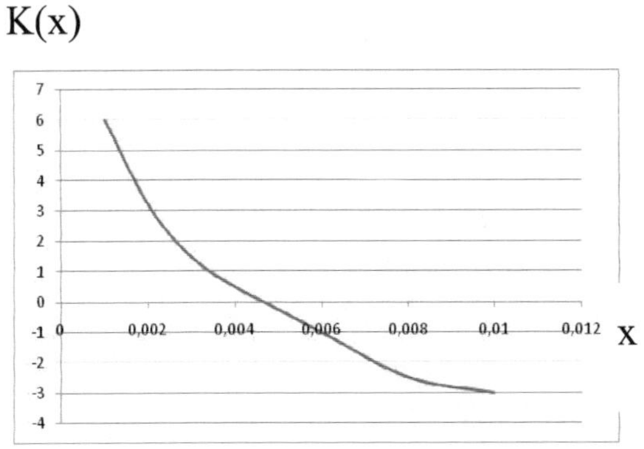

Figure 3: Distribution of the local heating in the inner regions of the RS Oph disk.

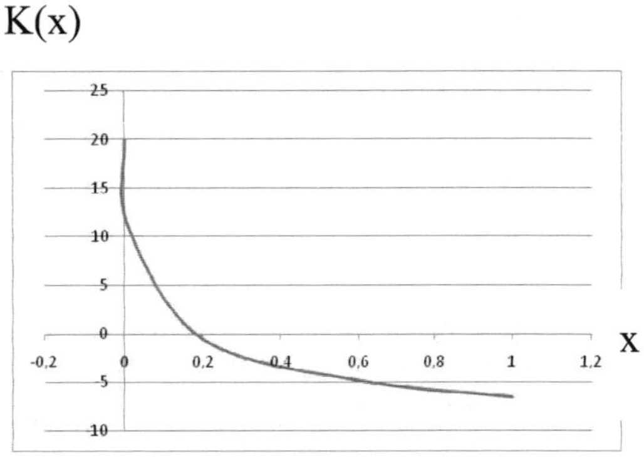

Figure 4: Distribution of the local heating in the Cyg X-1 disk.

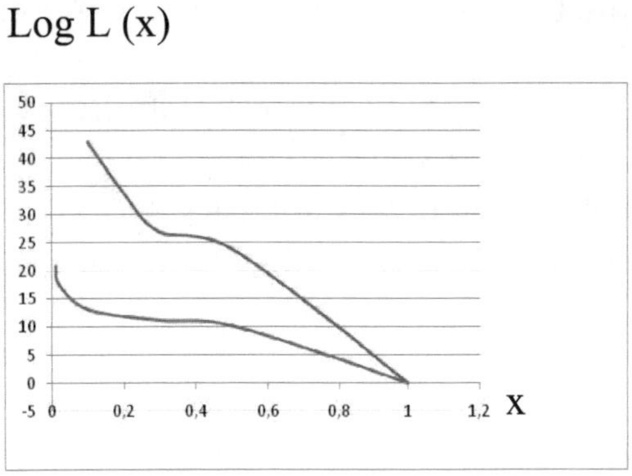

Figure 5: Luminosity distribution function of the RS Oph disk;

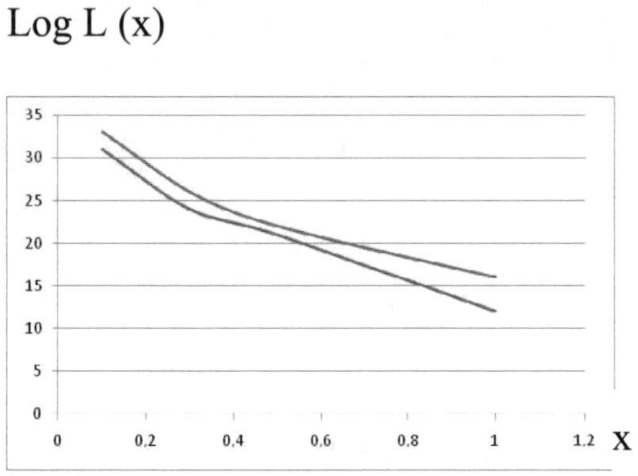

Figure 6: Luminosity distribution function of the Cyg X-1 disk

5. DISCUSSION

We can identify one main similarity and one subsequent, as well several significant differences in the active zones of our two objects:

1. The active zone consists of two separate parts – the rings in the inner and outer zone have different construction. The external ones have clearly expressed and separated from the environment. They keep in themselves the activity. While the internal rings pass one by other, they gradually fade, they blurred, which allows the activity to manifest itself.
2. In the early stages, the micro-quasar active zone is relatively narrower and heavily pulled inward to the accretor that in the disk of the other compact object.
3. In the active zone of RS Oph, however, the internal active zone is much smaller than the external one and the hidden activity is preserved in a much larger region of the disk.

The narrower and shifted in the inner regions active zone of Cyg X-1 is mainly due to the stronger background gravity of the black hole.

Weak gravity of the accretor allows the EM field in the disk more easily to hold the plasma in the rings (via partial freezing, the current efficiently guide the plasma). On this is due more than ten times longer outer active zone of RS Oph relative to the inner.

For comparison, in Cyg X-1 it is only two times larger.

4. At a later stage the active zones are extended in both objects:
5. At Cyg X-1, both active zones increase but retain location of their internal boundary.
6. At RS Oph the internal active zone increases and shortens the external active zone, equalizing approximately its width.

The pumped up by displacement activity in the inner layers is unfolding the inner zone in Cyg X-1. Here the powerful gravity plays a slightly different role, by her the accretor reacts, as unfolding and the outer zone, and thus preserving the inner boundary in the active zone.

At RS Oph the inner zone reacts in the same way, but the weaker gravity allows it to expand in both directions and to release the activity of the disk much sooner.

6. CONCLUSIONS

As the conclusion, we will note that the main structure, consequence of advection - the active zone and its construction in the various objects remain unchanged. This is a strong argument in support to the fundamental nature of the advection mechanism.

Differences in the active zones of the two disks are a result of the individual physical conditions caused by the specifics of the concrete objects and them evolution. This only confirms the functionality of our model at work with different types sources.

References

Beloborodov A.M.: 1999, *ASP Conf.Ser.,* 161, 295.

Fabian A.C., Wilkins D.R., Miller J.M., Reis R.C., Reynolds C.S., Cakett E.M., Nowak M.A., Pooley G.G., Pottshmidt K., Sanders J.S., Ross R.R., Wilms J.: 2012, *MNRAS*, 424, 217.

Giuseppe Frè P.: 2013, *Gravity, a Geometrical Course I*

Iankova Kr.D.: 2007, *BG-Ursi School and Workshop on Waves and Turbulence Phenomena in Space Plasmas, BSSPP Proceedings,* Series No. 1, 143-146, 2007, http://sp.phys.uni-sofia.bg/Kiten06/Pres/Iankova.pdf

Iankova Kr.: 2007, *Bulg. J. Phys.,* vol.34 no.s2 pp. 326-329.

Iankova Kr.D.: 2009, *Publ. Astron. Soc. "RudjerBošković"*, No. 9, 327-333, Belgrade.

Narayan, R., Yi, I.: 1994, *ApJ*, 428, *L13-L16.*

Narayan, R., Yi, I.: 1995, *ApJ*, 452, 71.

Yankova Kr.: 2012, *Publ. Astron. Soc. "RudjerBošković"*, vol. 11, 375-383.

Yankova Kr.: 2012, *JUBILEE INTERNATIONAL CONGRESS: SCIENCE, EDUCATION, TECHNOLOGIES "40 YEARS BULGARIA – SPACE COUNTRY ", proceedings (ISBN 978-954-577-636-6),* Tom1, 152-158.

Yankova Kr.D.: 2013, *Publ. Astron. Soc. "RudjerBošković"*, vol. 12, 375-381.

Yankova Kr.: 2014, *MHD of accretion-disk flows, (in Bulgarian).*

Yankova Kr.: 2015, *BlgAJ,* Vol. 22, 83.

Yankova, Kr.: 2015, *Proceedings Seventh Workshop "Solar Influences on the Magnetosphere, Ionosphere and Atmosphere",* 25-28, ISSN 2367-7570, http://ws-sozopol.stil.bas.bg/2015Sunny/Proceedings2015S1-2.pdf

Yankova, Kr.: 2015, *Publ. Astron. Soc. "RudjerBošković"*, No 15, 2015, 107-115.

Yankova, Kr.: 2017, *Proceedings Ninth Workshop "Solar Influences on the Magnetosphere, Ionosphere and Atmosphere",* 2017, 77-82, http://ws-sozopol.stil.bas.bg/2017Sunny/Proceedings2017_V3.pdf

Yankova, Kr.: 2017, *Proceedings SES 2017,* ISSN 1313-3888, 85-88.

Yankova Kr.: 2018, Fundamental Advection, *Proceedings SES2018*, 85-89, ISSN1313-3888.

Yankova Kr.: 2018, *Book of Proceedings Workshop "Solar Influences on the Magnetosphere, Ionosphere and Atmosphere",* SSTRI-BAN, 2018, ISSN:2367-7570, DOI:10.31401/WS.2018.proc, 43-46.

Yankova Kr.: 2019, *Proceedings Eleventh Workshop "Solar Influences on the Magnetosphere, Ionosphere and Atmosphere"*, 60-64, ISSN 2367-7570.

LIST OF PARTICIPANTS

AUTHORS' INDEX

PROGRAMME

LIST OF PARTICIPANTS

	Name	Affiliation	State
1	Alok Gupta	ARIES	IN
2	Anatoly Miroshnichenko	UNC	US
3	Asen Mutafov	IANAO	BG
4	Bojan Arbutina	UB	RS
5	Borislav Borisov	ShU	BG
6	Boyko Mihov	IANAO	BG
7	Daniela Boneva	SRTI	BG
8	Daniela Kirilova	IANAO	BG
9	Dejan Urosevic	UB	RS
10	Dimitar Sotirov	IANAO	BG
11	Dorotea Vasileva	ShU	BG
12	Dragomir Marchev	ShU	BG
13	Elena Vchkova-Bebekovska	US	NM
14	Evgeni Semkov	IANAO	BG
15	Gabriela Zidarova	ShU	BG
16	Galin Borisov	IANAO	BG
17	Goran Damljanovic	AOB	RS
18	Grozdan Shirov	IANAO	BG
19	Jakub Tokarek	AOI AMU	PL
20	Kiril Stoyanov	IANAO	BG
21	Krasimir Harizanov	ShU	BG
22	Krasimira Yankova	SRTI	BG
23	Krystian Ilkiewicz	AOUW	PL
24	Luba Dankova	IANAO	BG
25	Lubomir Iliev	IANAO	BG
26	Lyuba Slavcheva-Mihova	IANAO	BG
27	Magdalena Christova	TU	BG
28	Milan Dimitrijevic	AOB	RS
29	Milen Minev	SU	BG
30	Milica Vucetic	UB	RS
31	Miljana Jovanovic	AOB	RS
32	Miroslav Moyseev	IANAO, SU	BG
33	Momchil Dechev	IANAO	BG
34	Nataliya Pavlova	ShU	BG
35	Nikola Petrov	IANAO	BG
36	Ognyan Kounchev	IMI	BG
37	Orlin Stanchev	SU	BG

38	Petia Yanchulova	STSI	US
39	Renada Konstantinova-Antova	IANAO	BG
40	Rositsa Miteva	IANAO	BG
41	Rumen Bachev	IANAO	BG
42	Rumen Bogdanovski	IANAO	BG
43	Rumen Shkevov	SRTI	BG
44	Rumiana Bojilova	NIGGG	BG
45	Sava Donkov	IANAO	BG
46	Stanislav Milosevic	UB	RS
47	Stefan Stefanov	IANAO, SU	BG
48	Sunay Ibryamov	ShU	BG
49	Svetla Tsvetkova	IANAO	BG
50	Svetlana Boeva	IANAO	BG
51	Svetoslav Botev	SU	BG
52	Svetoslav Christov	INRNE	BG
53	Tanyu Bonev	IANAO	BG
54	Teodora Atanasova	ShU	BG
55	Todor Veltchev	SU	BG
56	Tsvetan Georgiev	IANAO	BG
57	Tsvetan Tsvetkov	IANAO	BG
58	Vladimir Bozhilov	SU	BG
59	Vladislav Marchev	IANAO	BG
60	Wojciech Dimitrov	AOI AMU	PL
61	Yavor Chapanov	CAWRI	BG
62	Yovelina Zinkova	IANAO	BG
63	Zoran Simic	AOB	RS

Affiliations:

BG - Bulgaria, **RS** - Serbia, **NM** – North Macedonia, **IN** – India, **PL** – Poland, **US** - USA

IANAO - Institute of Astronomy and NAO, Bulgarian Academy of Sciences
AOB - Astronomical Observatory of Belgrade
SU - Sofia University "St. Kliment Ohridski"
ShU - Konstantin Preslavsky University of Shumen
UB - University of Belgrade
IPhB - Institute of Physics Belgrade
SRTI - Space Research and Technology Institute, Bulgarian Academy of Sciences
IMI - Institute of Mathematics and Informatics, Bulgarian Academy of Sciences

TU - Technical University of Sofia

MGU - The University of Mining and Geology "St. Ivan Rilski"

CAWRI - The Climate, Atmosphere and Water Research Institute, Bulgarian Academy of Sciences

ARIES - Aryabhatta Research Institute of Observational Sciences, Nainital, India

UNC - University of North Carolina – Greensboro

AOUW - Astronomical Observatory of the University of Warsaw

AOI AMU - Astronomical Observatory Institute, Adam Mickiewicz University, Poznań

INRNE - The Institute for Nuclear Research and Nuclear Energy, Bulgarian Academy of Sciences

NIGGG - National Institute of Geophysics, Geodesy and Geography, Bulgarian Academy of Sciences

IICT - Institute of Information and Communication Technologies, Bulgarian Academy of Sciences

STSI - Space Telescope Science Institute

AUTHORS' INDEX

XIII BULGARIAN-SERBIAN ASTRONOMICAL CONFERENCE PROGRAMME

Monday 03.10.22

09:00-10:30 Registration and coffee break

10:30-10:40 Official opening

Evgeni Semkov, Milan Dimitrijević

10:40-12:30 Morning session. *Chair: Milan Dimitrijević*

10:40 – 11:00 Evgeni Semkov - The new 1.5-meter robotic telescope at the Rozhen Observatory

11:00 – 11:20 Rumen Bachev - Recent achievements at Belogradchik Observatory

11:20 – 11:40 Goran Damljanovic - Gaia Alerts and Bulgarian-Serbian cooperation from 2014 to 2022

11:40 – 12:10 Oliver Vince - Satus of Vidojevica 20 years after its founding (will be presented by Miljana Jovanovic)

12:10 – 12:30 Vladimir Bozhilov - The „Space Education for Bulgaria" (SpaceEdu4BG) Project

14:00-16:00 Afternoon session 1. *Chair: Evgeni Semkov*

14:00 – 14:30 Alok Gupta - Multi-wavelength variability and QPOs in Blazars

14:30 – 14:50 Miljana Jovanović - Comparison and control stars around quasars suitable for the ICRF – GAIA CRF link

14:50 – 15:10 Milen Minev - A photometric study of high-redshift quasars

15:10 – 15:40 Stanislav Milošević - Dark matter haloes in galaxy mergers

16:00-16:30 *Coffee break*

16:30-19:00 Afternoon session 2 *Chair: Evgeni Semkov*

16:30 – 17:00 Anatoly Miroshnichenko - B[e] star CI Cam in the optical range

17:00 – 17:30 Krystian Ilkiewicz - Flickering in polars

17:30 – 18:00 Svetla Tsvetkova - The large-scale magnetic fields of the M dwarf double-line spectroscopic binary FK Aqr

18:00-19:00 Poster session

18:00 Gabriela Zidarova - Long-term multicolor photometry of YSOs in the field of the open cluster Trumpler 37

18:03 Tsvetan Georgiev - Representation of asymmetric data distributions by intergrams instead histograms

18:06 Tsvetan Georgiev - Revealing of repeating structures in time series by structure length selection

18:09 Miroslav Moyseev - Photometric variability in star formation regions

18:12 Momchil Dechev - Bilateral Bulgarian-Indian project "Eruptions, flows and waves in solar atmosphere and their role in space weather": Overview of the results

18:15 Asen Mutafov - High-energy UV irradiance is dangerous to Earth's biological species

18:18 Zorica Cvetković - Bulgarian-Serbian Cooperation: CCD Observations of Visual Double Stars in the Period 2004-2022

18:21 Lyuba Slavcheva-Mihova - The application of the X-ray vs. UV luminosity relation for quasars

18:24 Boyko Mihov - Jet parameters for the blazar BL Lacertae

18:27 Daniela Boneva - On the disc's density distribution and temperature variations of two binary stars with compact objects

18:30 Goran Damljanovic - Processing double stars images using Machine Learning

18:33 Momchil Dechev - Different Type Prominence Eruptions Associated with CMEs

18:36 Jakub Tokarek – Searching for hierarchical systems in the Gaia Data Release 3

18:39 Rumiana Bojilova - Investigation of the sunrise effect in the ionosphere

Tuesday 04.10.22
09:30-11:00 Morning session 1 *Chair: Zoran Simić*

09:30 – 10:00 Sava Donkov - Thermodynamics of fluid elements in the context of saturated isothermal turbulence in Molecular Clouds

10:00 – 10:25 Milan Dimitrijević - Stark broadening of Fe XXV lines for neutron stars and their environment investigations

10:25 – 10:40 Magdalena Christova - Analysis of calculated Stark broadening parameters of singly ionized silicon lines

10:40 – 11:00 Svetoslav Botev - Properties of Milky way globular clusters associated with X-ray sources

11:00-11:30 *Coffee break*

11:30-12:50 Morning session 2 *Chair: Rositsa Miteva*

11:30 – 12:00 Bojan Arbutina - Low-mass Contact Close Binary Systems and Their Stability

12:00 – 12:20 Krasimira Yankova - The active zone mobility in a magnetized disk with advection

12:20 – 12:50 Dejan Urosevic - Supernovae throughout the history

14:30-16:00 Afternoon session 1 *Chair: Milica Vučetić*

14:30 – 15:00 Renada Konstantinova-Antova - Magnetic fields and activity in AGB and other evolved giant stars

15:00 – 15:20 Kiril Stoyanov - Optical spectroscopy and X-ray observations of the D-type symbiotic star EF Aql

15:20 – 15:40 Zoran Simić - Impact broadening parameters for some transition metals in A type star

15:40 – 16:00 Petia Yanchulova - Dust Extinction, 3D Structure, and Stellar Properties from Resolved Stars in Nearby Galaxies

16:00-16:30 *Coffee break*

16:30-18:10 Afternoon session 2 *Chair: Galin Borisov*

16:30 – 16:50 Lubomir Iliev - The carousel of cycles of a prototype Be/shell star Pleione

16:50 – 17:10 Svetlana Boeva - Negative superhumps in cataclysmic variables

17:10 – 17:30 Vladislav Marchev - For optical flickering in symbiotic star MWC 560

Wednesday 05.10.22

09:30-11:00 Morning session 1 *Chair: Lubomir Iliev*

10:00 – 10:20 Ognyan Kounchev - Generalized Smoothing splines for data analysis in Astronomy

10:20 – 10:40 Stefan Stefanov - The tilted accretion disc of LS Cam

10:40 – 11:10 Daniela Kirilova - Several cosmological nucleosynthesis constraints on neutrino and new physics

11:10-11:40 *Coffee break*

11:40-12:50 Morning session 2 *Chair: Dejan Urosevic*

11:40 – 12:10 Milica Vučetić - Supernova remnants - perspective from NAO Rozhen

12:10 – 12:50 Todor Veltchev - Power-law tails of the density distribution in star-forming clouds: possible effects of rotation and/or thermodynamics

12:50 – 13:00 Photo of conference participants

14:00-18:30 Excursion

19:30 Conference dinner

Thursday 06.10.22

09:30-11:00 Morning session 1 *Chair: Renada Konstantinova-Antova*

09:30 – 10:00 Rositsa Miteva - Solar radio bursts associated with in situ observed energetic protons and electrons.

10:00 – 10:30 Nikola Petrov - Optical observations of the solar chromosphere by NAO Rozhen.

10:30 – 10:40 Tsvetan Tsvetkov - Online Catalog of Solar Activity Events of Solar Cycle 24.

11:00-11:30 *Coffee break*

11:30-12:30 Morning session 2 *Chair Alok Gupta*

11:30 – 11:50 Yovelina Zinkova - Daytime sky polarization effect.

11:50 – 12:10 Galin Borisov - Physical and dynamical properties of selected Earth co-orbital asteroids.

12:10 – 12:30 Elena Vchkova-Bebekovska - Modeling asteroid shapes using BNAO Rozhen and AS Vidojevica photometric data in combination with sparse data.

14:30-16:00 Afternoon session 1 *Chair: Kiril Stoyanov*

14:30 – 14:50 Rumen Shkevov - Wave-particle resonant interactions in heliosphere periphery.

14:50 – 15:10 Yavor Chapanov - Wavelet Coherence of Total Solar Irradiance and Atlantic Climate.

15:10 – 15:40 Rumen G. Bogdanovski - INDIGO framework.

15:40 – 16:00 Grozdan Shirov - Analysis of meteorological observations for Bulgaria (1905-2022).

16:00-16:30 *Coffee break*

16:30-19:00 Afternoon session 2 *Chair: Kiril Stoyanov*

16:30 – 17:00 Tanyu Bonev - Mankind's first attempt to change the orbit of an asteroid.

17:40-19:00 Open discussion for future cooperation

CIP - Каталогизација у публикацији
Народна библиотека Србије, Београд

520/524(082)

SERBIAN-Bulgarian Astronomical Conference (13 ; 2022 ; Velingrad)
Proceedings of the XIII Serbian-Bulgarian Astronomical Conference, Velingrad, Bulgaria, October 3-7, 2022 / [Institute of Astronomy and National Astronomical Observatory, Bulgarian Academy of Sciences, Sofia, Bulgaria] ; [co-organizer Astronomical Observatory, Belgrade, Serbia] ; eds. Evgeni Semkov ... [et al.]. - Belgrade : Astronomical Society "Rudjer Bošković", 2023 (Beograd : Skripta Internacional). - 200 str. : ilustr. ; 24 cm. - (Публикације Астрономског друштва "Руђер Бошковић" ; sv. 25 = Publications of the Astronomical Society "Rudjer Bošković" ; no. 25)

Tiraž 100. - Napomene i bibliografske reference uz tekst. - Bibliografija uz svaki rad. - Registar.

а) Астрономија– Зборници б) Астрофизика – Зборници

COBISS.SR-ID 113537033